ISCN 2020

An International System for
Human Cytogenomic Nomenclature (2020)

Editors

Jean McGowan-Jordan, Ottawa, ON
Ros J. Hastings, Oxford
Sarah Moore, Adelaide, SA

Recommendations of the International Standing Committee on Human Cytogenomic Nomenclature Including Revised Sequence-Based Cytogenomic Nomenclature developed in collaboration with the Human Genome Variation Society (HGVS) Sequence Variant Description Working Group

10 figures, 4 tables, and foldout, 2020

Basel · Freiburg · Hartford · Oxford · Bangkok · Dubai · Kuala Lumpur · Melbourne · Mexico City · Moscow · New Delhi · Paris · Shanghai · Tokyo

Reprint of **Cytogenetic and Genome Research** (ISSN 1424–8581)
Vol. 160, No. 7–8, 2020

Copies of ISCN 2020 can be ordered from the publishers:

S. Karger AG	S. Karger Publishers, Inc.
Allschwilerstrasse 10	26 West Avon Road
P.O. Box	P.O. Box 529
CH–4009 Basel	Unionville, CT 06085
Switzerland	USA
Tel. +41 61 306 11 11	Tel. +1 860 675 7834
Fax +41 61 306 12 34	Fax +1 860 675 7302
E-Mail orders@karger.com	Toll free: +1 800 828 5479

Library of Congress Cataloging-in-Publication Data

Names: International Standing Committee on Human Cytogenomic Nomenclature, author. | McGowan-Jordan, Jean, editor. | Hastings, Ros J., editor. | Moore, Sarah (Cytogeneticist), editor.
Title: ISCN 2020 : an international system for human cytogenomic nomenclature (2020) / editor, Jean McGowan-Jordan, Ros J. Hastings, Sarah Moore.
Other titles: Cytogenetic and genome research.
Description: Basel ; Hartford : Karger, 2020. | „Recommendations of the International Standing Committee on Human Cytogenomic Nomenclature including revised sequence-based cytogenomic nomenclature developed in collaboration with the Human Genome Variation Society (HGVS) Sequence Variant Description Working Group." | Includes bibliographical references and index. | Summary: „An indispensable reference volume for human cytogeneticists, molecular geneticists, technicians, and students for the interpretation and communication of human cytogenetic and molecular cytogenomic nomenclature"-- Provided by publisher.
Identifiers: LCCN 2020040262 | ISBN 9783318067064 (alk. paper) | ISBN 9783318067040 (paperback ; alk. paper)
Subjects: MESH: Cytogenetics | Genomics | Terminology
Classification: LCC QH506 | NLM QU 15 | DDC 572.8--dc23
LC record available at https://lccn.loc.gov/2020040262

S. Karger
Medical and Scientific Publishers
Basel · Freiburg · Hartford · Oxford ·
Bangkok · Dubai · Kuala Lumpur ·
Melbourne · Mexico City ·
Moscow · New Delhi · Paris ·
Shanghai · Tokyo

Disclaimer
The statements, opinions and data contained in this publication are solely those of the individual authors and contributors and not of the publisher and the editor(s). The appearance of advertisements in the journal is not a warranty, endorsement, or approval of the products or services advertised or of their effectiveness, quality or safety. The publisher and the editor(s) disclaim responsibility for any injury to persons or property resulting from any ideas, methods, instructions or products referred to in the content or advertisements.

All rights reserved
No part of this publication may be translated into other languages, reproduced or utilized in any form or by any means, electronic or mechanical, including photocopying, recording, microcopying, or by any information storage and retrieval system, without permission in writing from the publisher.

© Copyright 2020 by S. Karger AG,
P.O. Box, CH–4009 Basel (Switzerland)
Printed on acid-free and non-aging paper (ISO 9706)
ISBN 978–3–318–06706–4
e-ISBN 978–3–318–06867–2

karger@karger.com
www.karger.com/cgr

The regular page numbers refer to Cytogenet Genome Res Vol. 160/7–8/2020
The page numbers in italics are used only in this reprint
Please always use the regular page numbers when citing ISCN 2020

Contents

1	**Historical Introduction**	**341|***1*
1.1	1956–1984	341|*1*
1.2	1985–1995	343|*3*
1.3	1996–2004	344|*4*
1.4	2005–2009	345|*5*
1.5	2010–2013	345|*5*
1.6	2014–2016	346|*6*
1.7	2017–2020	346|*6*
2	**Normal Chromosomes**	**347|***7*
2.1	Introduction	347|*7*
2.2	Chromosome Number and Morphology	347|*7*
2.2.1	Non-Banding Techniques	347|*7*
2.2.2	Banding Techniques	348|*8*
2.2.3	X- and Y-Chromatin	349|*9*
2.3	Chromosome Band Nomenclature	349|*9*
2.3.1	Identification and Definition of Chromosome Landmarks, Regions, and Bands	349|*9*
2.3.2	Designation of Regions, Bands, and Sub-Bands	351|*11*
2.4	High-Resolution Banding	352|*12*
2.5	Molecular Basis of Banding	354|*14*
3	**Symbols, Abbreviated Terms, and General Principles**	**374|***34*
4	**Karyotype Designation**	**378|***38*
4.1	General Principles	378|*38*
4.2	Specification of Breakpoints	381|*41*
4.3	Designating Structural Chromosome Aberrations by Breakpoints and Band Composition	382|*42*
4.3.1	Short Form for Designating Structural Chromosome Aberrations	382|*42*
4.3.1.1	Two-Break Rearrangements	382|*42*
4.3.1.2	Three-Break Rearrangements	382|*42*
4.3.1.3	Four-Break and More Complex Rearrangements	383|*43*
4.3.2	Detailed Form for Designating Structural Chromosome Aberrations	384|*44*
4.3.2.1	Additional Symbols	384|*44*
4.3.2.2	Designating the Band Composition of a Chromosome	384|*44*
4.4	Derivative Chromosomes	384|*44*
4.5	Recombinant Chromosomes	386|*46*

5	**Uncertainty in Chromosome or Band Designation**	**388|48**
5.1	Questionable Identification	388|48
5.2	Uncertain Breakpoint Localization or Chromosome Number	389|49
5.3	Alternative Interpretation	389|49
5.4	Incomplete Karyotype	390|50
6	**Order of Chromosome Abnormalities in the Karyotype**	**391|51**
7	**Normal Variable Chromosome Features**	**393|53**
7.1	Variation in Heterochromatic Segments, Satellite Stalks, and Satellites	393|53
7.1.1	Variation in Length	393|53
7.1.2	Variation in Number and Position	394|54
7.2	Fragile Sites	394|54
8	**Numerical Chromosome Abnormalities**	**395|55**
8.1	General Principles	395|55
8.2	Sex Chromosome Abnormalities	396|56
8.3	Autosomal Abnormalities	397|57
9	**Structural Chromosome Rearrangements**	**399|59**
9.1	General Principles	399|59
9.2	Specification of Structural Rearrangements	400|60
9.2.1	Additional Material of Unknown Origin	400|60
9.2.2	Deletions	401|61
9.2.3	Derivative Chromosomes	402|62
9.2.4	Dicentric Chromosomes	407|67
9.2.5	Duplications	409|69
9.2.6	Fission	409|69
9.2.7	Fragile Sites	410|70
9.2.8	Homogeneously Staining Regions	410|70
9.2.9	Insertions	411|71
9.2.10	Inversions	412|72
9.2.11	Isochromosomes	412|72
9.2.12	Marker Chromosomes	413|73
9.2.13	Neocentromeres	415|75
9.2.14	Quadruplications	415|75
9.2.15	Ring Chromosomes	415|75
9.2.16	Telomeric Associations	417|77
9.2.17	Translocations	418|78
9.2.17.1	Reciprocal Translocations	418|78
9.2.17.2	Whole-Arm Translocations	420|80
9.2.17.3	Robertsonian Translocations	421|81
9.2.17.4	Jumping Translocations	422|82
9.2.18	Tricentric Chromosomes	422|82
9.2.19	Triplications	423|83
9.3	Multiple Copies of Rearranged Chromosomes	423|83

10	**Chromosome Breakage**	**425|85**
10.1	Chromatid Aberrations	425|85
10.1.1	Non-Banded Preparations	425|85
10.1.2	Banded Preparations	426|86
10.2	Chromosome Aberrations	426|86
10.2.1	Non-Banded Preparations	426|86
10.2.2	Banded Preparations	427|87
10.3	Scoring of Aberrations	427|87
11	**Neoplasia**	**428|88**
11.1	Clones and Clonal Evolution	428|88
11.1.1	Definition of a Clone	428|88
11.1.2	Clone Size	429|89
11.1.3	Mainline	429|89
11.1.4	Stemline, Sideline and Clonal Evolution	430|90
11.1.5	Composite Karyotype	432|92
11.1.6	Unrelated Clones	434|94
11.2	Modal Number	434|94
11.3	Constitutional Karyotype	435|95
11.4	Counting Chromosome Aberrations	437|97
12	**Meiotic Chromosomes**	**438|98**
12.1	Terminology	438|98
12.1.1	Examples of Meiotic Nomenclature	439|99
12.1.2	Correlation between Meiotic Chromosomes and Mitotic Banding Patterns	441|101
13	***In situ* Hybridization**	**446|106**
13.1	Introduction	446|106
13.2	Prophase/Metaphase *in situ* Hybridization (ish)	447|107
13.2.1	Normal Signal Pattern	447|107
13.2.2	Abnormal Signal Patterns with Single Probes	448|108
13.2.3	Abnormal Signal Patterns with Multiple Probes	449|109
13.2.4	Abnormal Mosaic and Chimeric Signal Patterns with Single or Multiple Probes	452|112
13.2.5	Oncology-Specific Exceptions where Multiple Copies of the Same Gene Are Present	453|113
13.2.6	Use of dim and enh in Metaphase *in situ* Hybridization	454|114
13.2.7	Subtelomeric Metaphase *in situ* Hybridization	454|114
13.3	Interphase/Nuclear *in situ* Hybridization (nuc ish)	455|115
13.3.1	Number of Signals	455|115
13.3.2	Normal Interphase Signal Pattern	456|116
13.3.3	Abnormal Interphase Signal Pattern	456|116
13.3.4	Donor versus Recipient	458|118
13.3.5	Relative Position of Signals	458|118
13.3.5.1	Single Fusion Probes	461|121
13.3.5.2	Single Fusion with Extra Signal Probes	461|121
13.3.5.3	Dual Fusion Probes	461|121
13.3.5.4	Break-Apart Probes	461|121
13.3.5.5	Tricolor Probes	462|122

13.4	*In situ* Hybridization on Extended Chromatin/DNA Fibers (fib ish)	463|*123*
13.5	Reverse *in situ* Hybridization (rev ish)	463|*123*
13.5.1	Chromosome Analyses Using Probes Derived from Sorted or Microdissected Chromosomes	463|*123*
13.6	Multi-Color Chromosome Painting	464|*124*
13.7	Partial Chromosome Paints	464|*124*
14	**Microarrays**	**465|*125***
14.1	Introduction	465|*125*
14.2	Examples of Microarray Nomenclature	466|*126*
14.2.1	Normal	466|*126*
14.2.2	Abnormal	466|*126*
14.2.3	Inheritance	469|*129*
14.2.4	Multiple Techniques	470|*130*
14.2.5	Mixed Cell Populations and Uncertain Copy Number	472|*132*
14.2.6	Nomenclature Specific to SNP Array	474|*134*
14.2.7	Complex Array Results	476|*136*
14.2.8	Polar Bodies	477|*137*
15	**Region-Specific Assays**	**480|*140***
15.1	Introduction	480|*140*
15.2	Examples of RSA Nomenclature for Normal and Aneuploidy	480|*140*
15.3	Examples of RSA Nomenclature for Partial Gain or Loss	481|*141*
15.4	Examples of RSA Nomenclature for Balanced Translocations or Fusion Genes	483|*143*
16	**Sequence-Based Nomenclature for Description of Chromosome Rearrangements**	**484|*144***
16.1	Introduction	484|*144*
16.2	General Principles	484|*144*
16.3	Examples of Sequence-Based Nomenclature for Description of Large Structural Variation	486|*146*
16.3.1	Deletions	486|*146*
16.3.2	Derivative Chromosomes	487|*147*
16.3.3	Duplications	488|*148*
16.3.4	Insertions	488|*148*
16.3.5	Inversions	489|*149*
16.3.6	Ring Chromosomes	489|*149*
16.3.7	Translocations	490|*150*
16.4	Examples of Sequence-Based Nomenclature for Description of Large Copy Number Variation	491|*151*
17	**References**	**492|*152***
18	**Members of the ISCN Standing Committee**	**495|*155***
19	**Appendix**	**497|*157***
20	**Index**	**499|*159***
	The Normal Human Karyotype G- and R-bands	**after Index**

1 Historical Introduction

1.1 1956–1984[1]

In 1956 Tjio and Levan, in their now classic article, reported that the human chromosome number was 46 and not 48. This work, which was carried out on cultured human embryonic cells, was rapidly confirmed by studies of testicular material by Ford and Hamerton (1956). These two articles stimulated a renewed interest in human cytogenetics, and, by 1959, several laboratories were engaged in the study of human chromosomes and a variety of classification and nomenclature systems had been proposed. This resulted in confusion in the literature and a need to establish a common system of nomenclature that would improve communication between workers in the field.

For this reason, a small study group was convened in Denver, Colorado at the suggestion of Charles E. Ford. Fourteen investigators and three consultants participated, representing each of the laboratories that had published human karyotypes up to that time. The system proposed in the report of this meeting, entitled "A Proposed Standard System of Nomenclature of Human Mitotic Chromosomes," more commonly known as the Denver Conference (1960), has formed the basis for all subsequent nomenclature reports and has remained virtually unaltered, despite the rapid developments of the last 25 years. It is fair to say that the participants at Denver did their job so well that this report has formed the cornerstone of human cytogenetics since 1960, and the foresight and cooperation shown by these investigators have prevented much of the nomenclature confusion which has marked other areas of human genetics.

Three years later, a meeting called by Lionel S. Penrose was held in London (London Conference, 1963) to consider developments since the Denver Conference. The most significant result of that conference was to give official sanction to the classification of the seven groups of chromosomes by the letters A to G, as originally proposed by Patau (1960).

The next significant development came in Chicago at the Third International Congress on Human Genetics in 1966 when 37 investigators, representing the major cytogenetic laboratories, met to determine whether it was possible to improve the nomenclature and thus eliminate some of the major problems that had resulted from the rapid proliferation of new findings since 1960. The report of this conference (Chicago Conference, 1966) proposed a standard system of nomenclature for the provision of short-hand descriptions of the human chromosome complement and its abnormalities, a system that, in its basic form, has stood

---

[1] Adapted from ISCN (1985).

karger@karger.com
www.karger.com/cgr

© 2020 S. Karger AG, Basel

the test of time and is now used throughout the world for the description of non-banded chromosomes.

In his introductory address to the Chicago Conference (1966), Lionel Penrose made the following prophetic statement:

It is easy to be carried away by the detectable peculiarities and to forget that much underlying variability is still hidden from view until some new technical device discloses the finer structure of chromosomes, as in the Drosophila salivary gland cells.

Two years later in 1968, the second major breakthrough occurred when Torbjörn Caspersson and his colleagues, working in Sweden, published the first banding pictures of plant chromosomes stained with quinacrine dihydrochloride or quinacrine mustard (Caspersson et al., 1968). These studies were rapidly expanded to human chromosomes by these workers, who published the first banded human karyotype in 1970 (for a review of this work, see Caspersson et al., 1972). Soon, several other techniques that also produced chromosome bands were developed. This led to the realization that, as each human chromosome could now be identified very precisely, the existing system of nomenclature would no longer be adequate.

A group of 50 workers concerned with human cytogenetics met in 1971 on the occasion of the Fourth International Congress of Human Genetics in Paris to agree upon a uniform system of human chromosome identification. Their objective was accomplished and extended by the appointment of a Standing Committee, chaired by John Hamerton, which met initially in Edinburgh in January 1972, and then with a number of expert consultants at Lake Placid in New York in December 1974, and again in Edinburgh in April 1975.

The 1971 meeting in Paris, together with the 1972 Edinburgh meeting of the Standing Committee, resulted in the report of the Paris Conference (1971), a highly significant document in the annals of human cytogenetics. This document proposed the basic system for designating not only individual chromosomes but also chromosome regions and bands, and it provided a way in which structural rearrangements and variants could be described in terms of their band composition.

By 1974 it had become clear that the number of workers in the field was now too great to allow the holding of such conferences as the Chicago and Paris ones, where the majority of laboratories involved could be represented. The Standing Committee therefore proposed holding smaller, nonrepresentative conferences, each on a number of fairly specific topics and that would utilize expert consultants for each topic. The first meeting of this type was held in 1974 in Lake Placid and the second in 1975 in Edinburgh, at which a number of specific topics – including heteromorphic chromosomes of the Hominoidea, and chromosome registers – were discussed. These discussions were reported in the 1975 supplement to the Paris Conference report (Paris Conference, 1971, Supplement, 1975).

A further change came about in 1976 at the Fifth International Congress of Human Genetics in Mexico City, when a meeting of all interested human cytogeneticists was held to elect an International Standing Committee on Human Cytogenetic Nomenclature. These elections provided a truly international and geographic representation for the Standing Committee and provided a mandate to the committee to continue its work in proposing ways in which human chromosome nomenclature might be improved. Jan Lindsten was appointed the chairman of this committee.

The committee met in Stockholm in 1977 and, following past practice, invited a number of expert consultants to meet with it. It was decided at this meeting to cease labeling reports geographically and to unify the various conference reports reviewed above into a document entitled "An International System for Human Cytogenetic Nomenclature (1978)," to be abbreviated ISCN (1978). ISCN (1978) included all major decisions of the Denver, London, Chicago, and Paris Conferences, without any major changes but edited for consistency and accuracy. It thus provided in one document a complete system of human cytogenetic nomenclature that has stood the test of time and has proved to be of value not only to those entering the field for the first time but also to experienced cytogeneticists.

The next major area to be considered by the Standing Committee was the nomenclature of chromosomes stained to show "high resolution banding." In 1977 a working party was established under the chairmanship of Bernard Dutrillaux to consider this matter.

It had been recognised for some time that prophase and prometaphase chromosomes reveal a much larger number of bands than can be seen even in the best banded metaphase chromosome preparations. Techniques were devised to partially synchronise peripheral blood cultures so as to yield sufficient cells in the early phase of mitosis for detailed study. These all essentially use some method of blocking cells in the S-phase, releasing the block and then timing the subsequent harvest to obtain the maximum number of cells at the appropriate stage (Dutrillaux, 1975; Yunis, 1976). Several studies showed that techniques of this kind required a new nomenclature (Francke and Oliver, 1978; Viegas-Pequignot and Dutrillaux, 1978; Yunis et al., 1978).

The working group met on several occasions. There was a remarkable degree of agreement on the number of bands, the width of the bands and their relative positions. There was, however, considerable difficulty in reaching a consensus on the origin of certain bands and on the stage of their appearance relative to other bands. A broad measure of agreement was, however, reached at a meeting in Paris in May 1980 and this was published as "An International System for Human Cytogenetic Nomenclature – High Resolution Banding (1981)" or ISCN (1981).

A new Standing Committee was elected at a specially convened meeting of cytogeneticists held during the Sixth International Congress of Human Genetics in Jerusalem in 1981. David Harnden was appointed chairman of the new committee.

A revision of the International System for Human Cytogenetic Nomenclature was prepared in 1984, to be published as ISCN (1985), partly because a reprint was in any case necessary and partly because, once again, it was felt to be important to try to keep all statements on nomenclature together in a single volume. The opportunity was taken to correct errors and make a small number of amendments but no attempt was made to make a major revision.

The widely accepted international nomenclature for human chromosomes has proved to be an important element in improving and maintaining international collaboration. The development of this system has been made possible by the collaboration of many people. I would like to thank not only members of the Standing Committee but others who have acted as consultants or who have contributed ideas or materials to these publications. In particular I would like to express the gratitude of the international cytogenetic community to the March of Dimes Birth Defects Foundation for its consistent and substantial financial support over the past 19 years. Without its help none of these developments would have been possible.

David Harnden
October 1984

1.2   1985–1995

A new Standing Committee was elected at a meeting of cytogeneticists attending the Seventh Congress of Human Genetics held in Berlin in 1986 and Uta Francke was appointed as chairman. The Committee was aware of a considerable increase in the amount and variety of data on chromosome aberrations associated with neoplasia, and considered that a terminology was necessary for those acquired chromosome aberrations that were not adequately described by the nomenclature for constitutional aberrations as published in ISCN (1985). A subcommittee under the chairmanship of Felix Mitelman was established and charged with the task of producing a nomenclature for cancer cytogenetics. The report of this subcommittee was adopted by the ISCN Standing Committee and published as "ISCN (1991): Guidelines for Cancer Cytogenetics." These guidelines superseded previous ISCN recommendations on cancer cytogenetics and have since come into general use.

A new Standing Committee was elected at the Eighth International Congress of Human Genetics held in Washington, DC, in 1991 and Felix Mitelman was appointed chairman. The new committee considered that it would be timely to review and update the ISCN (1985) nomenclature in the light of developments in the field, including advances in the use of *in situ* hybridization techniques, and to incorporate all revisions and the guidelines for cancer cytogenetics into a single document to be published as ISCN (1995). Cytogeneticists were asked, through notices published in relevant journals, to forward to the Committee their comments on any defects in the ISCN 1978–1991 publications, as well as any suggestions for alterations and improvements.

The Standing Committee and consultants met in Memphis on October 9–13, 1994, at the kind invitation of Professor Avirachan Tharapel. The Committee considered all the recommendations that had been submitted to it and updated, modified, and amalgamated the 1985 and 1991 documents into a single text with the intention of this being published in 1995.

H.J. Evans
P.A. Jacobs
October 1994

1.3   1996–2004

The Ninth International Congress of Human Genetics was held in Rio de Janeiro in 1996. A new Standing Committee was elected at the satellite meeting of the cytogeneticists. Patricia A. Jacobs was appointed as the chairperson of the Committee. In light of the extensive revision of ISCN (1995), the new Committee elected not to implement additional changes during its term.

The Tenth International Congress of Human Genetics was held in Vienna. The congregation of cytogeneticists present at the satellite meeting elected a new committee and Niels Tommerup was appointed as chairman. The extensive use of ISCN (1995) by the scientific community identified several areas that needed clarifications, deletions and additions. Therefore, the Committee decided to review and update the ISCN (1995). The seven members of the Committee and nine external consultants met in Vancouver, BC, December 8–10, 2004 at the invitation of Niels Tommerup and Lisa G. Shaffer. The primary changes included replacing G- and R-banded karyotypes (Figs. 2 and 3) with new ones reflecting higher band-level resolutions, the addition of a new idiogram at the 300-band level, and introduction of a new 700-band-level idiogram that reflected the actual size and position of bands. The *in situ* hybridization nomenclature was modernized, simplified, and expanded. New examples reflecting unique situations were added, and a basic nomenclature for recording array comparative genomic hybridization results was introduced.

The Committee adopted changes to its membership structure for the future. The number of members was expanded to eleven from the current seven to reflect better representation of the geographic distribution of cytogeneticists. The voting constituency and guidelines for the election of members and chairpersons was redefined. Lisa Shaffer was appointed as chairman of the newly elected Committee. Finally, the Committee recommended that ISCN (2005) be published in 2005.

D.H. Ledbetter
A.T. Tharapel
December 2004

1.4 2005–2009

Early in 2006, Lisa Shaffer and Niels Tommerup organized the election for the next Standing Committee. Ballots were distributed and collected worldwide, and at the Eleventh International Congress of Human Genetics, held in Brisbane, Australia, in 2006, the results of the election were announced, resulting in a Committee of eleven elected members. The newly elected Committee received feedback on ISCN (2005) and decided to hold a meeting in 2008 to discuss potential changes and additions to a new edition of ISCN. At the invitation of Lisa Shaffer, Chair, the Committee and two external consultants met in Vancouver, BC, October 8–10, 2008. The primary change in cancer was the accommodation for either **idem** or **sl/sdl** in the nomenclature to describe clonal evolution. The *in situ* hybridization nomenclature was further clarified and additional examples provided. The basic microarray nomenclature was revised and expanded to accommodate all platform types, with more examples provided. Finally, a nomenclature for MLPA was introduced. The Committee recommended that ISCN (2009) be published in 2009.

Lisa G. Shaffer
Marilyn L. Slovak
Lynda J. Campbell
December 2008

1.5 2010–2013

In the fall of 2011, Lisa Shaffer organized the election for the next Standing Committee. The Committee was reduced to eight members including three from the Americas, three from Europe, one from Asia and one from Africa/Australia/New Zealand/Oceania. Ballots were distributed and collected worldwide, and the results of the election were announced. The newly elected Committee received feedback on ISCN (2009) and decided to hold a meeting in the spring of 2012 to discuss potential changes and additions to a new edition of ISCN. At the invitation of Lisa Shaffer, Chair, the Committee and two external consultants met in Seattle, Washington, April 10–11, 2012. During the meeting, Jean McGowan-Jordan was elected as the new Chair of the ISCN Committee. The Committee spent substantial time discussing that the primary purpose of the ISCN is to foster communication among cytogeneticists using a standard nomenclature that can be used to describe any genomic rearrangement identified either by standard karyotyping or molecular methodologies. The primary changes to the new edition of ISCN include additional illustrative examples of uses of nomenclature, inclusion of some definitions including chromothripsis and duplication, and the use of the genome build when describing microarray results. In ISCN (2009) MLPA nomenclature was introduced. The Committee considered adding nomenclature for other targeted quantitative assays such as QF-PCR, real-time-PCR and bead-based multiplex techniques, but decided to delete section 14.4 on MLPA and rather introduce a new chapter 15 for nomenclature that can be used for any Region-Specific Assay (RSA). Finally, the Committee decided to delete any symbols that are not used in the nomenclature. With these changes, the Committee recommended that ISCN (2013) be published.

Lisa G. Shaffer
Jean McGowan-Jordan
May 2012

1.6 2014–2016

In the spring of 2014, Jean McGowan-Jordan, as Chair, contacted the members of the Standing Committee regarding the accumulating need to describe chromosomal abnormalities identified by sequence-based technologies. The relatively brief period since the publication of ISCN 2013 and the willingness of Committee members to maintain their commitment facilitated a meeting of the Committee in San Diego, Calif., October 22–23, 2014, which included two invited advisors. Prior to the meeting, input on required changes and corrections to ISCN 2013 was sought from the Cytogenetics community. Various approaches to describing chromosome abnormalities characterized by DNA sequencing were considered and discussed during a special joint session with the members of the Human Genome Variation Society (HGVS) Sequence Variant Description Working Group. Due to the long-standing use of HGVS terms and rules for description of sequence-based changes, it was decided that the HGVS and ISCN would collaborate on the development of a new nomenclature which would work for both the Molecular Genetics and Cytogenetics communities. It was agreed that this new scheme would form a new chapter of ISCN 2016. The Committee agreed upon required corrections and changes and the addition of new examples, particularly for microarray and region-specific assays, including the requirement to incorporate the genome build in the HGVS-standard format whenever nucleotide numbers are specified. Changes in the main text compared to the previous edition would be marked in the margin for the convenience of the reader. The decision to modify the name of the Standing Committee and nomenclature scheme to reflect changes in technology under its purview, by incorporating the term "Cytogenomic" (as replacement for Cytogenetic), was also made. The Committee then recommended the publication of ISCN 2016, An International System for Human Cytogenomic Nomenclature (2016).

Jean McGowan-Jordan
Annet Simons
December 2015

1.7 2017–2020

In the fall of 2018, Jean McGowan-Jordan, as Chair, organized the nomination and election of three new members for the next Standing Committee. Input on required changes and corrections to ISCN 2016 was sought from the Cytogenomics community. A meeting of the Committee occurred in Gothenburg, Sweden, June 13–14, 2019. Due to the increased use of technologies such as microarray and sequencing which orientate chromosomes by nucleotide number from pter to qter, the Committee decided to standardize this approach across all technologies, including banded chromosomes. The resolution to standardize the presentation of sex chromosome abnormalities before those affecting autosomes for all technologies was also made. It was decided to adopt a nomenclature for inherited abnormalities that clarified whether the rearrangement is inherited intact or partially as a derivative. The Committee identified the needs for specific nomenclature to the analysis of polar bodies, and to improve the existing nomenclature based on sequencing technology. The changes in the main text compared to the previous ISCN 2016 version are marked in the margin to assist the reader. With these improvements, the Committee recommended the publication of ISCN 2020.

Jean McGowan-Jordan
Ros J. Hastings
April 2020

2 Normal Chromosomes

2.1 Introduction

Human chromosome nomenclature is based on the results of several international conferences (Denver 1960, London 1963, Chicago 1966, Paris 1971, Paris 1975, Stockholm 1977, Paris 1980, Memphis 1994, Vancouver 2004, Vancouver 2008, Seattle 2012, San Diego 2014, Gothenburg 2019). The present report, which summarizes the current nomenclature, incorporates and supersedes all previous ISCN recommendations. The ISCN Standing Committee recommends that this nomenclature system be used also in other species.

2.2 Chromosome Number and Morphology

2.2.1 Non-Banding Techniques

In the construction of the karyogram[2] the autosomes are numbered from 1 to 22 in order of decreasing length (one exception is that chromosome 21 is shorter than chromosome 22). The sex chromosomes are referred to as X and Y.

When the chromosomes are stained by methods that do not produce bands, they can be arranged into seven readily distinguishable groups (A–G) based on descending order of size and the position of the centromere.

The group letter designations placed before the chromosome numbers are those agreed upon at the London Conference (1963). Not all chromosomes in the D and G groups show satellites on their short arms in a single cell. The number and size of these structures are variable.

The following parameters were used to describe non-banded chromosomes: (1) the length of each chromosome, expressed as a percentage of the total length of a normal haploid set, i.e., the sum of the lengths of the 22 autosomes and of the X chromosome; (2) the arm ratio of the chromosomes, expressed as the length of the longer arm relative to the shorter one;

---

[2] The terms *karyogram, karyotype,* and *idiogram* have often been used indiscriminately. The term *karyogram* should be applied to a systematized array of the chromosomes prepared either by drawing, digitized imaging, or by photography, with the extension in meaning that the chromosomes of a single cell can typify the chromosomes of an individual or even a species. The term *karyotype* should be used to describe the normal or abnormal, constitutional or acquired, chromosomal complement of an individual, tissue or cell line. We recommend that the term *idiogram* be reserved for the diagrammatic representation of a karyotype.

and (3) the centromeric index, expressed as the ratio of the length of the shorter arm to the whole length of the chromosome. The latter two indices are, of course, related algebraically.

Group A (1–3)	Large metacentric chromosomes readily distinguished from each other by size and centromere position.
Group B (4–5)	Large submetacentric chromosomes.
Group C (6–12, X)	Medium-sized metacentric or submetacentric chromosomes. The X chromosome resembles the longer chromosomes in this group.
Group D (13–15)	Medium-sized acrocentric chromosomes with satellites.
Group E (16–18)	Relatively short metacentric or submetacentric chromosomes.
Group F (19–20)	Short metacentric chromosomes.
Group G (21–22, Y)	Short acrocentric chromosomes with satellites. The Y chromosome bears no satellites.

The definition of metacentric, submetacentric, and acrocentric traditionally relates to the ratio of the chromosome arms in unbanded preparations (Levan et al., 1964; Al-Aish, 1969).

2.2.2 Banding Techniques

Numerous technical procedures have been reported that produce banding patterns on metaphase chromosomes.

A **band** is defined as the part of a chromosome that is clearly distinguishable from its adjacent segments by appearing darker or lighter with one or more banding techniques. Bands that stain darkly with one method may stain lightly with other methods. The chromosomes are visualized as consisting of a continuous series of light and dark bands, so that, by definition, there are no "interbands."

The methods first published for demonstrating bands along the chromosomes were those that used quinacrine mustard or quinacrine dihydrochloride to produce a fluorescent banding pattern. These methods are named Q-staining methods and the resulting bands **Q-bands** (Fig. 1). The numbers assigned to each chromosome were based on the Q-banding pattern as given by Caspersson et al. (1972). Techniques that demonstrate an almost identical pattern of dark and light bands along the chromosomes usually use the Giemsa dye mixture as the staining agent. These techniques are generally termed G-staining methods and the resulting bands **G-bands** (Fig. 2). Some banding techniques give patterns that are opposite in staining intensity to those obtained by the G-staining methods, viz, the reverse staining methods, and the resulting bands are called **R-bands** (Fig. 3).

The banding techniques fall into two principle groups: (1) those resulting in bands distributed along the length of the whole chromosome, such as G-, Q-, and R-bands, including techniques that demonstrate patterns of DNA replication, and (2) those that stain specific chromosome structures and hence give rise to a restricted number of bands (Table 1). These include methods that reveal constitutive heterochromatin (**C-bands**) (Fig. 4), telomeric bands (**T-bands**), and nucleolus organizing regions (**NORs**). For the code to describe banding techniques, see Table 2.

The patterns obtained with the various C-banding methods do not permit identification of every chromosome in the somatic cell complement but, as demonstrated in Table 1, can be used to identify specific chromosomes. The C-bands on chromosomes 1, 9, 16, and Y are all morphologically variable. The short-arm regions of the acrocentric chromosomes also demonstrate variations in size and staining intensity of the Q-, G-, R-, C-, T-, and NOR-bands. These variations are heritable features of the particular chromosome.

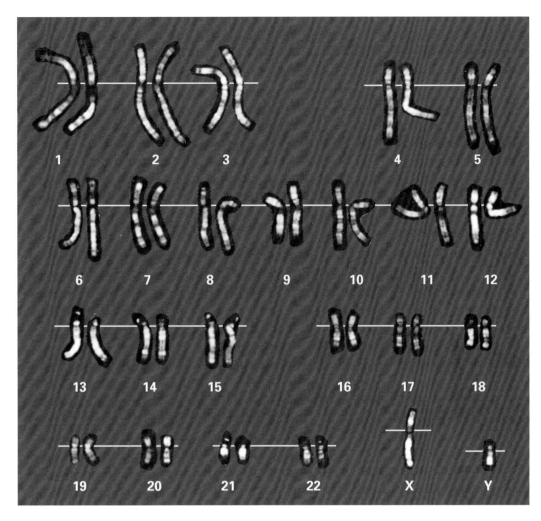

Fig. 1. Q-banded human karyogram. (Courtesy of Dr. E. Magenis.)

2.2.3 X- and Y-Chromatin

Inactive X chromosomes, as well as the heterochromatic segment on the long arm of the Y chromosome, appear as distinctive structures in interphase nuclei, for which the terms **X-chromatin** (Barr body, sex chromatin, X-body) and **Y-chromatin** (Y-body), respectively, should be used.

2.3 Chromosome Band Nomenclature

2.3.1 Identification and Definition of Chromosome Landmarks, Regions, and Bands

Each chromosome in the human somatic cell complement is considered to consist of a continuous series of bands, with no unbanded areas. As defined earlier, a **band** is a part of a chromosome clearly distinguishable from adjacent parts by virtue of its lighter or darker staining intensity. The bands are allocated to various regions along the chromosome arms,

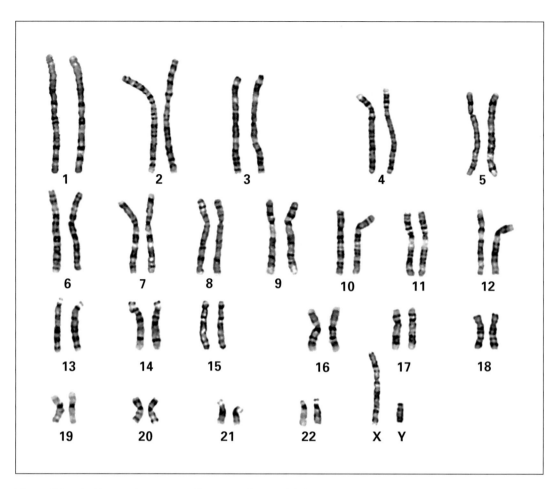

Fig. 2. G-banded human karyogram. (Courtesy of N.L. Chia.)

and the regions are delimited by specific **landmarks**. These are defined as consistent and distinct morphologic features important in identifying chromosomes. Landmarks include the ends of the chromosome arms, the centromere, and certain bands. The bands and the regions are numbered from the centromere outward. A **region** is defined as an area of a chromosome lying between two adjacent landmarks.

The original banding pattern was described in the Paris Conference (1971) report and was based on the patterns observed in different cells stained with either the Q-, G-, or R-banding technique (Appendix, Chapter 19). The banding patterns obtained with these staining methods agreed sufficiently to allow the construction of a single diagram representative of all three techniques. The bands were designated on the basis of their midpoints and not by their margins. Intensity was taken into consideration in determining which bands should serve as landmarks on each chromosome in order to divide the chromosome into natural, easily recognizable morphologic regions. A list of bands serving as landmarks that were used in constructing this diagram is provided in Table 3.

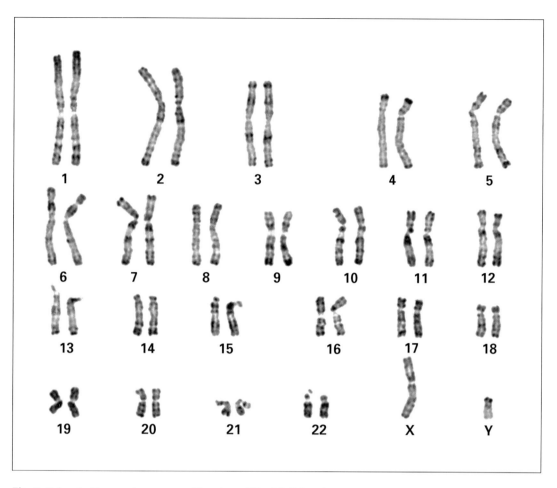

Fig. 3. R-banded human karyogram. (Courtesy of Dr. M. Prieur.)

2.3.2 Designation of Regions, Bands, and Sub-Bands

Regions and bands are numbered consecutively from the centromere outward along each chromosome arm. The symbols **p** and **q** are used to designate, respectively, the short and long arms of each chromosome. The centromere (**cen**) itself is designated 10; the part facing the short arm is p10, the part facing the long arm is q10. These are not shown in the idiograms. The two regions adjacent to the centromere are labeled as 1 in each arm; the next, more distal regions as 2, and so on. A band used as a landmark is considered as belonging entirely to the region distal to the landmark and is accorded the band number of 1 in that region.

In designating a particular band, four items are required: (1) the chromosome number, (2) the arm symbol, (3) the region number, and (4) the band number within that region. These items are given in order without spacing or punctuation. For example, 1p31 indicates chromosome 1, short arm, region 3, band 1.

Whenever an existing band is subdivided, a decimal point is placed after the original band designation and is followed by the number assigned to each sub-band. The sub-bands are numbered sequentially from the centromere outward. For example, if the original band 1p31 is subdivided into three equal or unequal sub-bands, the sub-bands are labeled 1p31.1, 1p31.2, and 1p31.3, sub-band 1p31.1 being proximal and 1p31.3 distal to the centromere.

Table 1. Examples of heteromorphisms with various stains[a.]

Technique	Chromosome									
	1	2	3	4	5	6	7	8	9	10
G[b]	q12 inv(p13q21)	inv(p11.2q13)	inv(p11.2q12)			p11.1			q12 inv(p12q13)	inv(p11.2q21.2)
C[c]	qh								qh	
G11	qh		cen		q11.1		p11.1		qh	q11.1
R or T	p36.3	q37		p16	p15.3 q35		p22	q24.3	q34	q26
NOR										
Q[d]			cen		cen					
DA-DAPI[e]	qh								qh	

Technique	Chromosome												
	11	12	13	14	15	16	17	18	19	20	21	22	X Y
G[b]			p	p	p	q11.2 inv(p11.2q12.1)					p	p	inv(p11.2q11.2)
C[c]			p	p	p	qh	p11				p	p	q12
G11			p	p	p		p11.1				q11.1	p	q12
R or T	p15 q13	p13	p12 q34	p12 q32	p12	p13.3 q24	q25			p13.3 q13.1	q13	p12 q22	p12 q11.2 q13
NOR			p12	p12	p12						p12	p12	
Q[d]			p11.2 p13 cen	p11.2 p13	p11.2 p13						p11.2 p13	p11.2 p13	
DA-DAPI[e]					p11.2	qh							q12

[a] cen = Centromere, h = heterochromatin, inv = inversion, p = short arm, q = long arm.
[b] Only the most commonly seen heteromorphisms are listed.
[c] All centromeres show constitutive heterochromatin variation.
[d] Only the brilliant and intensity-variable Q-bands are listed.
[e] DA-DAPI = Distamycin A and 4′,6-diamidino-2-phenylindole.

If a sub-band is subdivided, additional digits, but no further punctuation, are used; e.g., sub-band 1p31.1 might be further subdivided into 1p31.11, 1p31.12, etc. Although in principle a band can be subdivided into any number of new bands at any one stage, a band is usually subdivided into three sub-bands.

2.4 High-Resolution Banding

The nomenclature for high-resolution preparations of prophase and prometaphase chromosomes set forth by ISCN (1981) is an extension of the nomenclature for the banding patterns for metaphase chromosomes established at the Paris Conference (1971) and in ISCN (1978). The original system was specifically devised to allow for expansion as more chromosome bands were recognized.

High-resolution banding techniques can be applied to chromosomes in different stages of the cell cycle, e.g., prophase, prometaphase, or interphase (by methods that induce premature chromosome condensation). Furthermore, the number of discernible bands depends not only on the state of condensation but also on the banding technique used. The level of resolution is determined by the number of bands seen in a haploid set (22 autosomes + X and Y). The

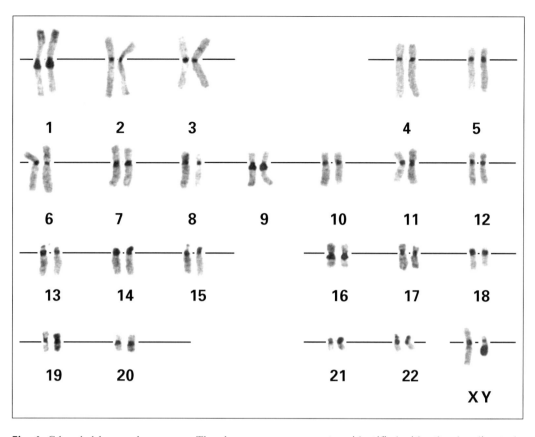

Fig. 4. C-banded human karyogram. The chromosomes were not preidentified with other banding techniques. (Courtesy of Dr. N. Mandahl.)

standard idiograms shown in Fig. 5 provide schematic representations of chromosomes corresponding to approximately 300, 400, 550, 700 and 850 bands. Although larger numbers of bands can be visualized, 550- to 850-band idiograms are sufficient for practical purposes. The 400- and 550-band idiograms are taken from ISCN (1985), and the 850-band idiogram was introduced in ISCN (1981) (Francke, 1994). The original nomenclature was based on patterns rather than on measurements. Also, variation in intensity of staining, which is dependent on the staining technique, was not reflected in the ISCN (1981) idiograms. At higher resolution, however, an idiogram depicting only patterns of black and white bands becomes difficult to use. Therefore, the ISCN (1981) 850-band idiogram has been replaced by previously published idiograms that are based on measurements of trypsin-Giemsa bands on prometaphase chromosomes and include five different shades of staining intensities to facilitate orientation among the large number of bands (Francke, 1981, 1994).

The ISCN (2005) added the 300- and 700-band idiograms as additional references. The idiograms, which show a G-band pattern, are provided to represent the position of bands in G-, Q- and R-stained preparations. Although the appearance of bands visualized by G-staining may differ from that revealed by R-staining (Fig. 6), the sequence of the bands is the same.

Two kinds of variable regions are indicated by different cross-hatching patterns, one involving the pericentromeric heterochromatin regions on all chromosomes and the other involving the variable regions 1q12, 3q11.2, 9q12, 16q11.2, 19p12, 19q12, Yq12, and the short arms of all acrocentric chromosomes. The representations of these variable regions are not

Table 2. Examples of the code used to describe banding techniques. In this one-, two-, or three-letter code, the first letter denotes the type of banding, the second letter the general technique, and the third letter the stain.

Q	Q-bands
QF	Q-bands by fluorescence
QFQ	Q-bands by fluorescence using quinacrine
QFH	Q-bands by fluorescence using Hoechst 33258
G	G-bands
GT	G-bands by trypsin
GTG	G-bands by trypsin using Giemsa
GTL	G-bands by trypsin using Leishman
GTW	G-bands by trypsin using Wright
GAG	G-bands by acetic saline using Giemsa
C	C-bands
CB	C-bands by barium hydroxide
CBG	C-bands by barium hydroxide using Giemsa
R	R-bands
RF	R-bands by fluorescence
RFA	R-bands by fluorescence using acridine orange
RH	R-bands by heating
RHG	R-bands by heating using Giemsa
RB	R-bands by BrdU
RBG	R-bands by BrdU using Giemsa
RBA	R-bands by BrdU using acridine orange
DA-DAPI	DAPI-bands by Distamycin A and 4′,6-diamidino-2-phenylindole

based on measurements. Banded structures can be seen within the variable regions, in particular in 1q12, 9q12 and Yq12, but since they are variable they have not been detailed in the idiograms. Normal chromosome variants are discussed in more detail in Chapter 7.

The lowest band number of 10 is assigned to the centromere (not shown on idiograms). The adjacent heterochromatic regions carry band designations of 11, 11.1 or 11.11 depending on the level of resolution.

One problem in assigning numbers to euchromatic sub-bands is that in G-banded preparations new G-bands appear to arise by subdivision of darkly stained G-bands on less extended chromosomes, while in R-staining preparations the dark R-bands appear to split. These interpretations of band to sub-band relationships would lead to different number assignments. Therefore, in assigning sub-band numbers, arbitrary decisions were made for the purposes of nomenclature only that should not be interpreted as statements about chromosome physiology. Examples of G- and R-banded chromosomes at successive stages of resolution are shown in Fig. 6a and b. In addition, G- and R-banded metaphase chromosomes at approximately the 550-band level and their diagrammatic representation (modified from ISCN 1985) are illustrated in a detachable foldout on the inside of the backcover.

2.5 Molecular Basis of Banding

Chromosome bands reflect the functional organization of the genome that regulates DNA replication, repair, transcription, and genetic recombination. The bands are large structures, each approximately 5 to 10 megabases of DNA that may include hundreds of genes. The molecular basis of banding methods is known to involve nucleotide base composition, associated proteins, and genome functional organization. In general, Giemsa-positive bands

Table 3. Bands serving as landmarks that divide the chromosomes into cytologically defined regions. The omission of an entire chromosome or chromosome arm indicates that either both arms or the arm in question consists of only one region, delimited by the centromere and the end of the chromosome arm.

Chromosome Number	Arm	Number of regions	Landmarks[a]
1	p	3	Proximal band of medium intensity (21), median band of medium intensity (31)
	q	4	Proximal negative band (21) distal to variable region, median intense band (31), distal band of medium intensity (41)
2	p	2	Median negative band (21)
	q	3	Proximal negative band (21), distal negative band (31)
3	p	2	Median negative band (21)
	q	2	Median negative band (21)
4	q	3	Proximal negative band (21), distal negative band (31)
5	q	3	Median band of medium intensity (21), distal negative band (31)
6	p	2	Median negative band (21)
	q	2	Median negative band (21)
7	p	2	Distal band of medium intensity (21)
	q	3	Proximal band of medium intensity (21), median band of medium intensity (31)
8	p	2	Median negative band (21)
	q	2	Median band of medium intensity (21)
9	p	2	Median intense band (21)
	q	3	Median band of medium intensity (21), distal band of medium intensity (31)
10	q	2	Proximal intense band (21)
11	q	2	Median negative band (21)
12	q	2	Median band of medium intensity (21)
13	q	3	Median intense band (21), distal intense band (31)
14	q	3	Proximal intense band (21), distal band of medium intensity (31)
15	q	2	Median intense band (21)
16	q	2	Median band of medium intensity (21)
17	q	2	Proximal negative band (21)
18	q	2	Median negative band (21)
21	q	2	Median intense band (21)
X	p	2	Proximal band of medium intensity (21)
	q	2	Proximal band of medium intensity (21)

[a] The numbers in parentheses are the region and band numbers as shown in Fig. 5.

(G-dark bands, R-light bands) are AT-rich, late replicating, and gene poor; whereas, Giemsa-negative bands (G-light bands, R-dark bands) are CG-rich, early replicating, and relatively gene rich.

Centromeric DNA and pericentromeric heterochromatin, composed of α-repetitive DNA and various families of repetitive satellite DNA, are easily detected by C-banding. The telomere is composed of 5 to 20 kb of tandem hexanucleotide minisatellite repeat units, TTAGGG, and stains darkly by T-banding. The 18S and 28S ribosomal RNA genes are clustered together in large arrays containing about 40 copies of each gene. These are located on the acrocentric short arms, at the nucleolar organizer regions or NORs, and are detected by silver staining.

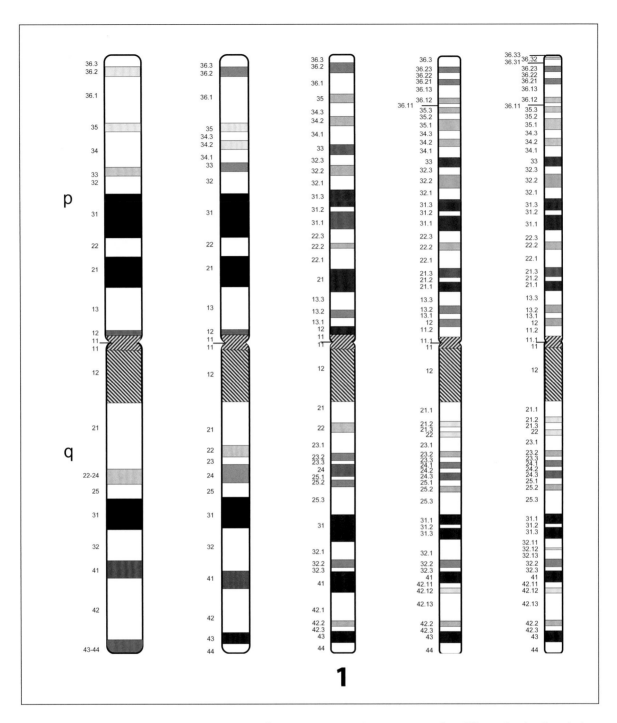

Fig. 5. Idiograms of G-banding patterns for normal human chromosomes at five different levels of resolution. From the left, chromosomes in each group represent a haploid karyotype of approximately 300-, 400-, 550-, 700-, and 850-band levels. The location and width of bands are not based on any measurements. The dark G-bands correspond to bright Q-bands, with the exception of the variable regions. The numbering of R-banded chromosomes is exactly the same, with a reversal of light and dark bands. While the band numbers are exactly the same, the relative widths of euchromatic bands are based on measurements and the staining intensities re-

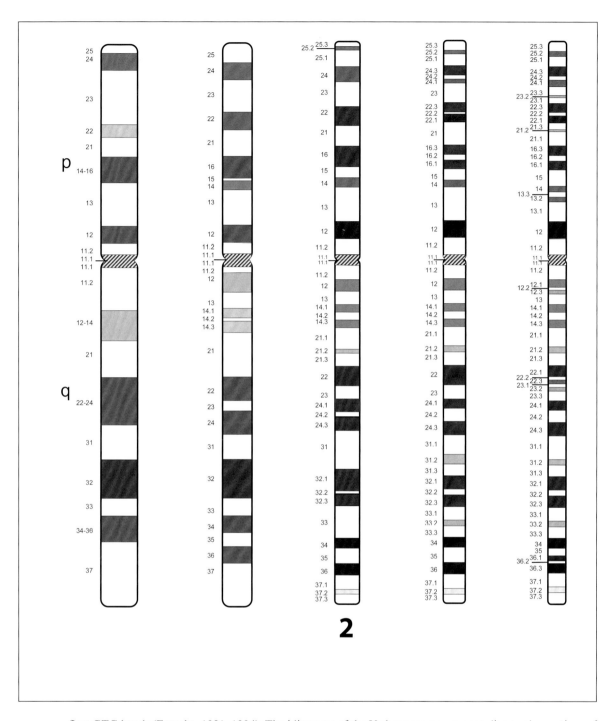

flect GTG bands (Francke, 1981, 1994). The idiograms of the Y chromosome are according to observations of Magenis and Barton (1987). While the number of bands on the euchromatic portion of the long arm has been expanded, the designations for light versus dark bands have been maintained. The 400-, 550-, and 850-band idiograms correspond to the ISCN (1995) nomenclature. The 300- and 700-band idiograms, new to ISCN (2005), were provided by N.L. Chia.

Normal Chromosomes

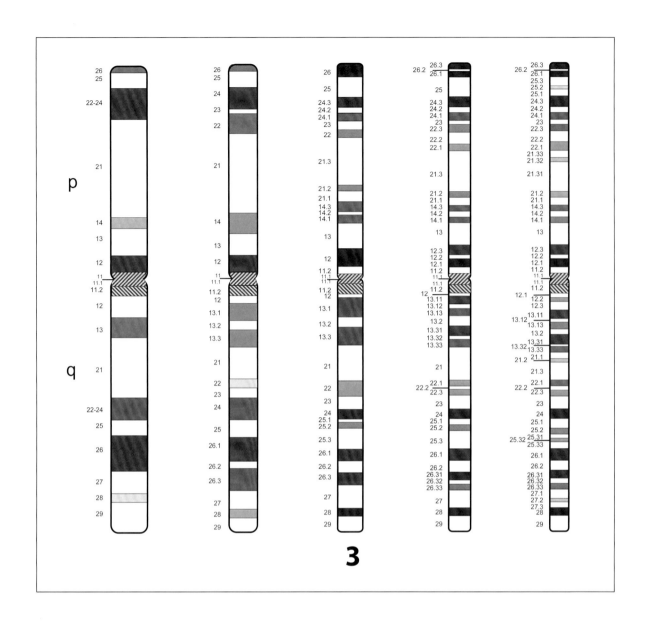

Fig. 5. continued (see legend on pp 356–357)

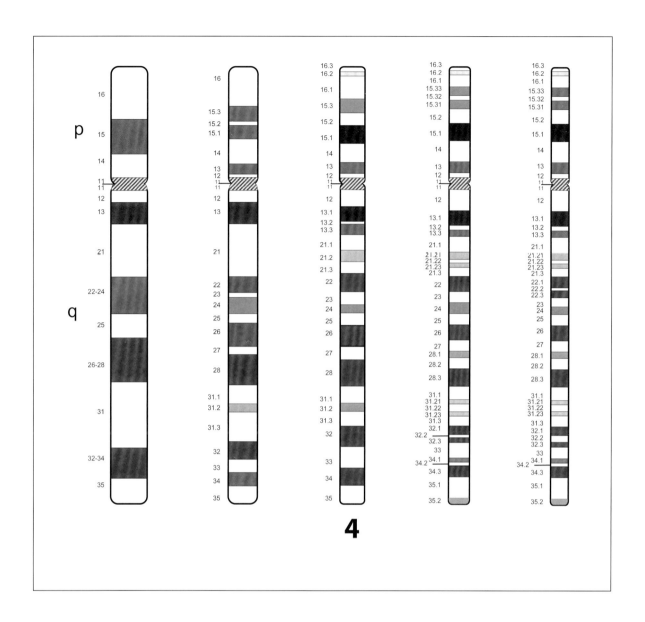

Normal Chromosomes

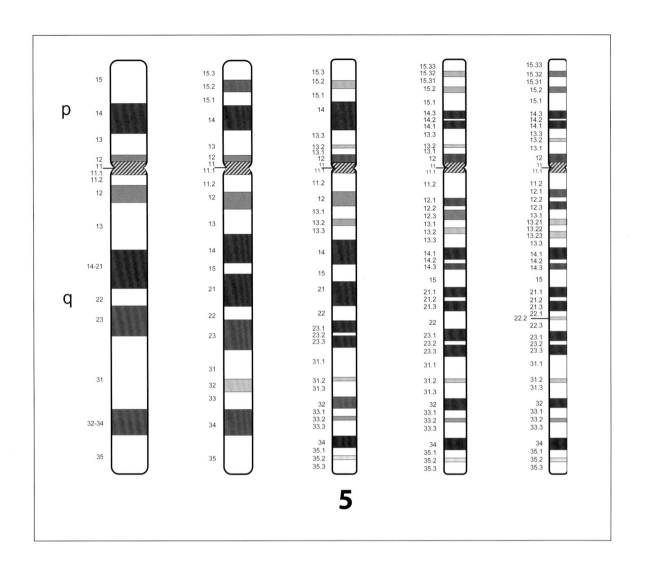

Fig. 5. continued (see legend on pp 356–357)

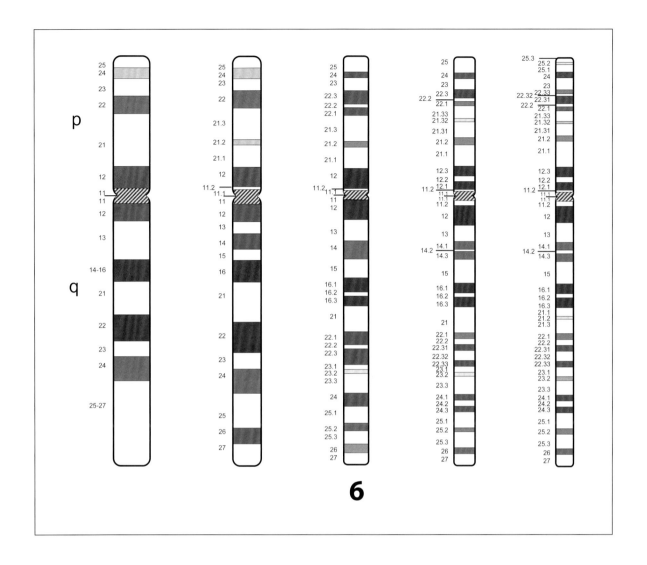

Normal Chromosomes

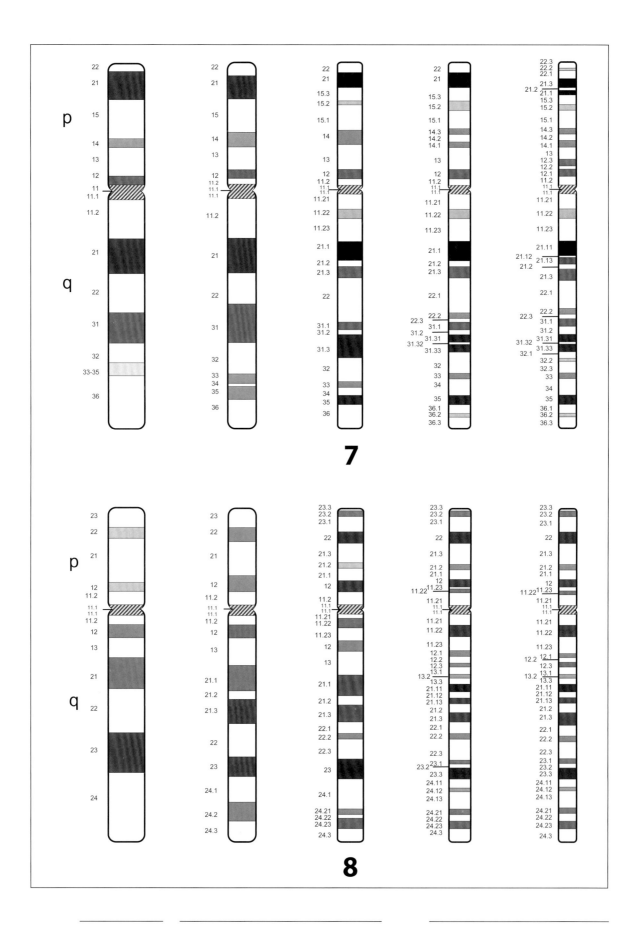

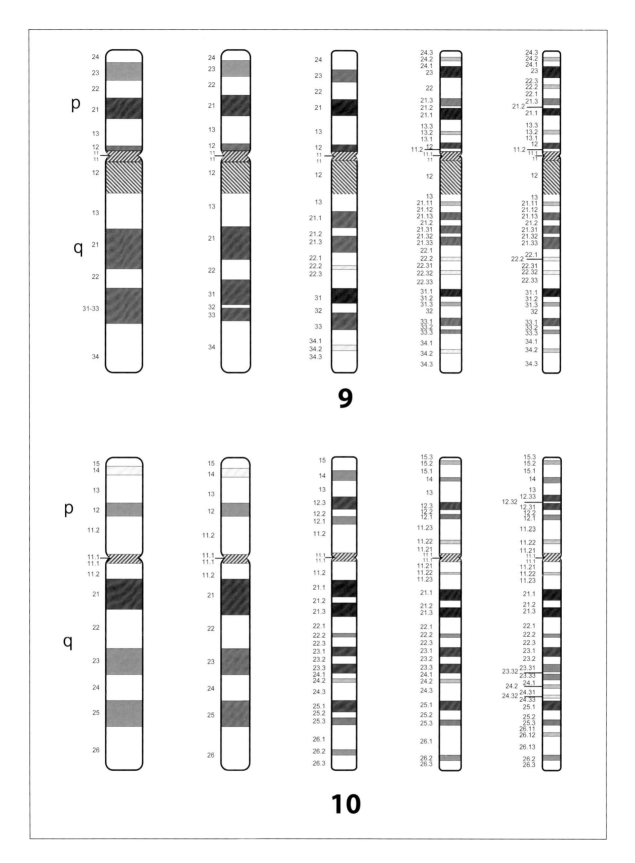

Fig. 5. continued (see legend on pp 356–357)

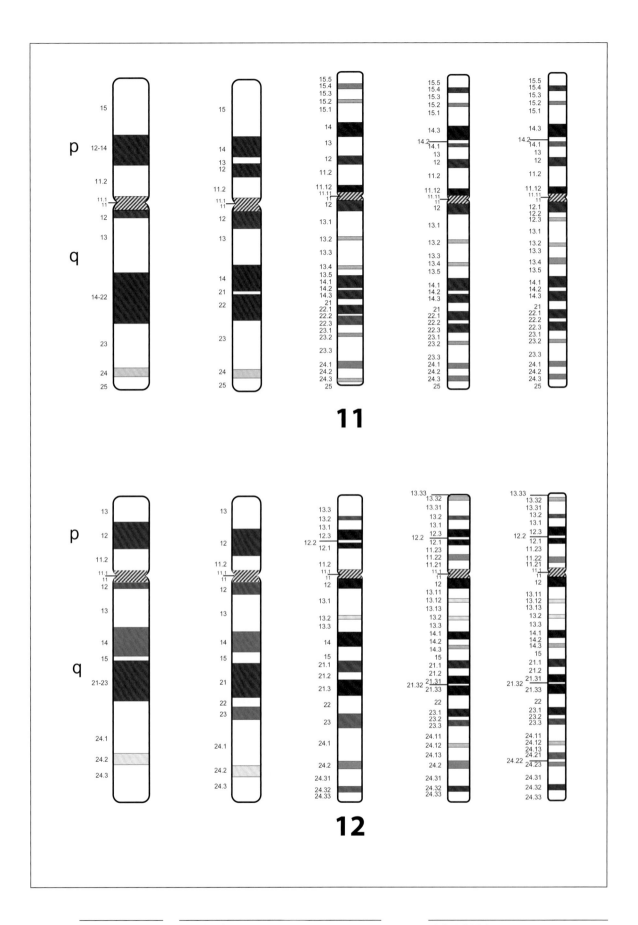

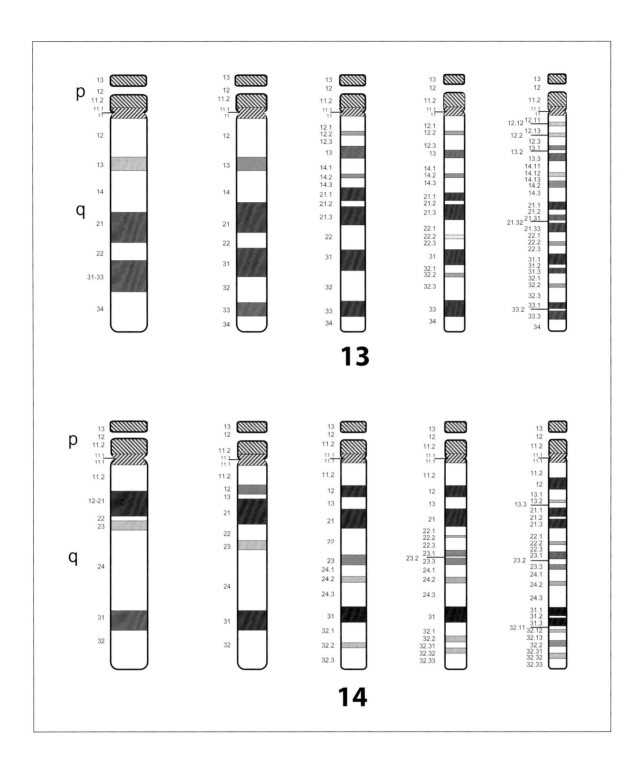

Fig. 5. continued (see legend on pp 356–357)

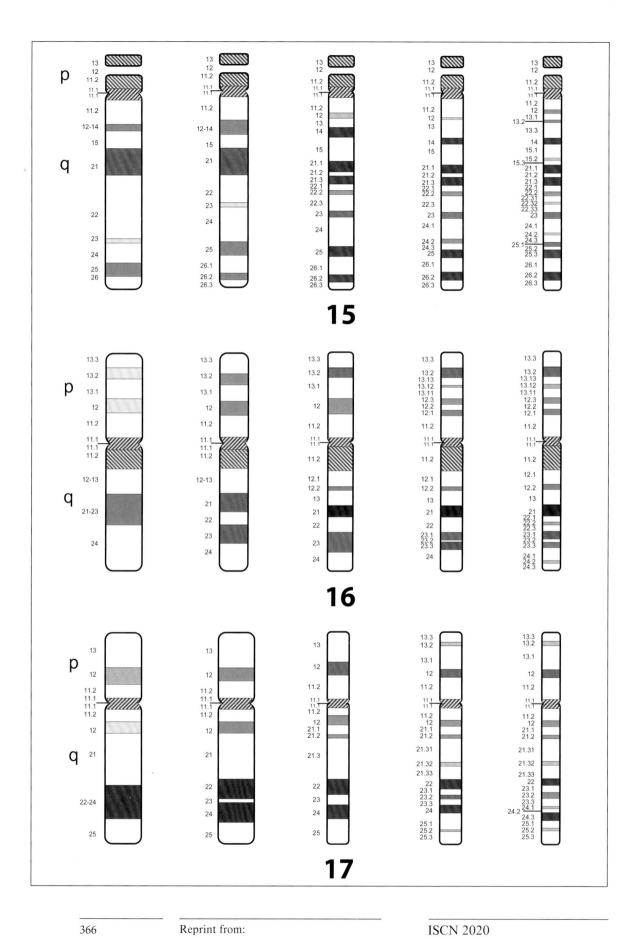

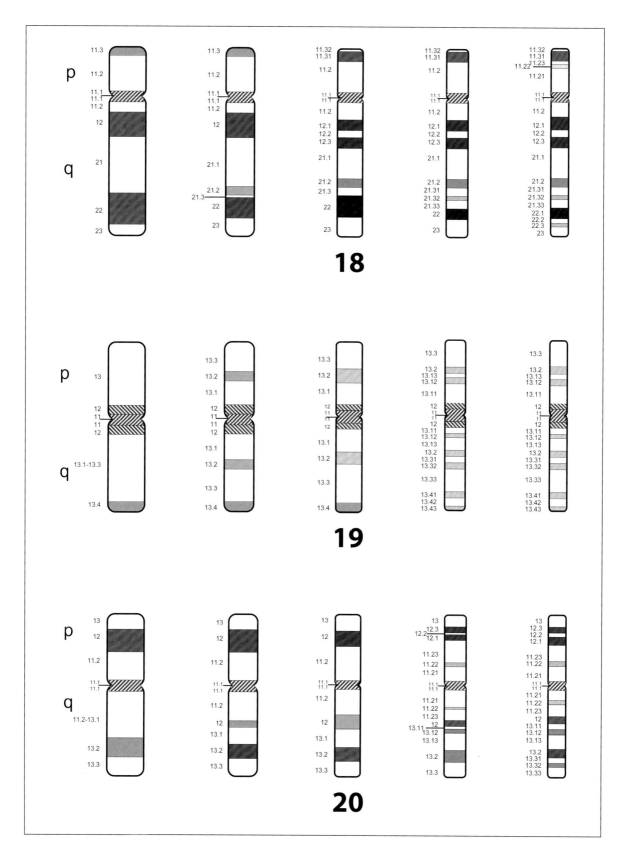

Fig. 5. continued (see legend on pp 356–357)

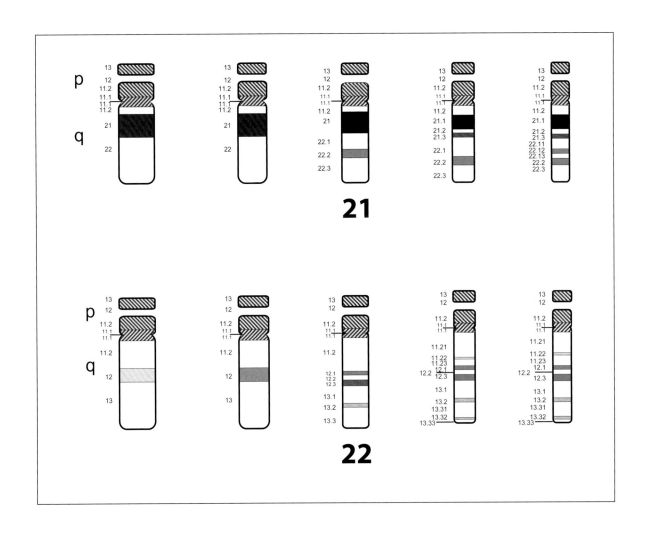

Fig. 5. continued (see legend on pp 356–357)

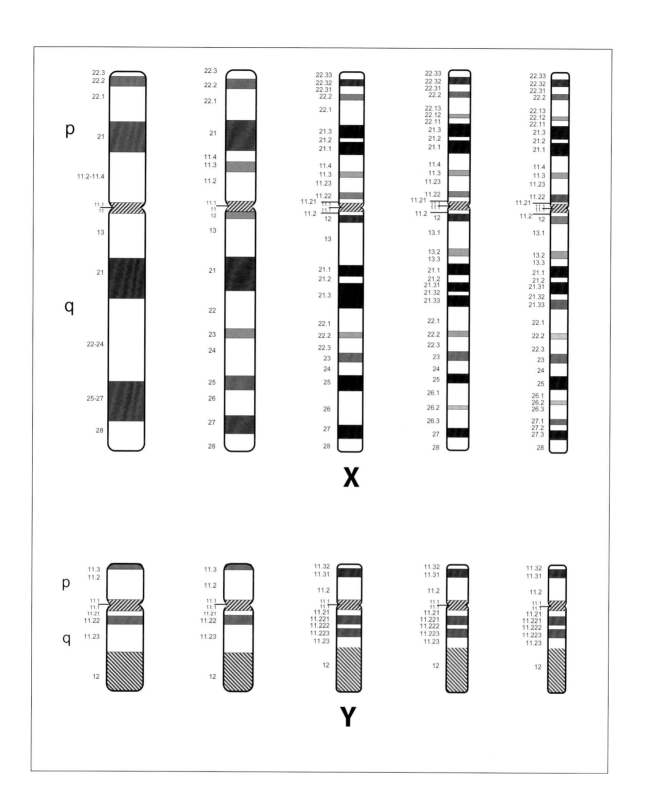

Normal Chromosomes

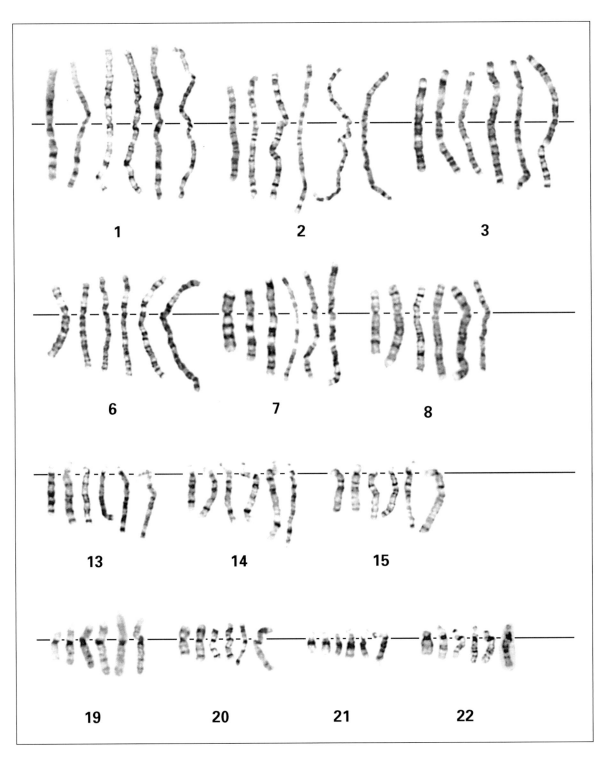

Fig. 6. (a) G-banded chromosomes arranged in increasing order of resolution from approximately the 500- to the 900-band levels. (Courtesy of Dr. E. Magenis.)

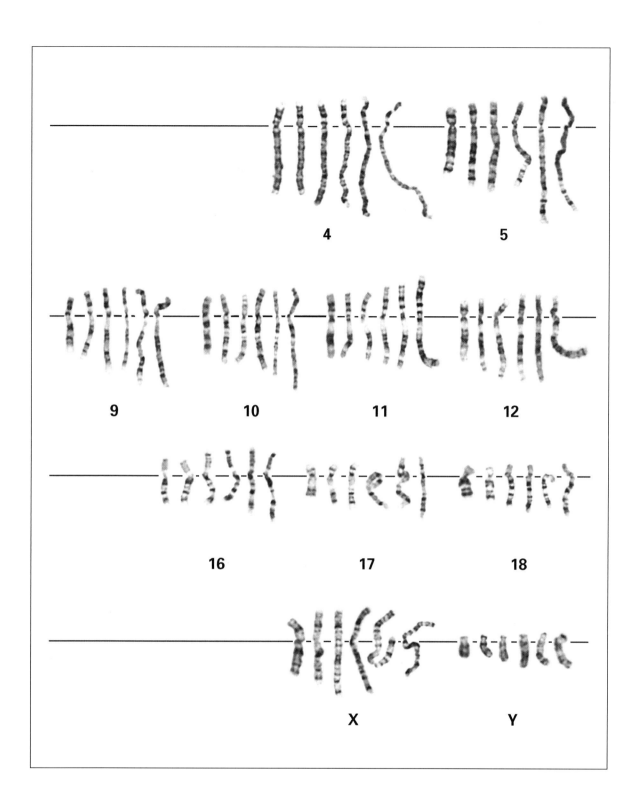

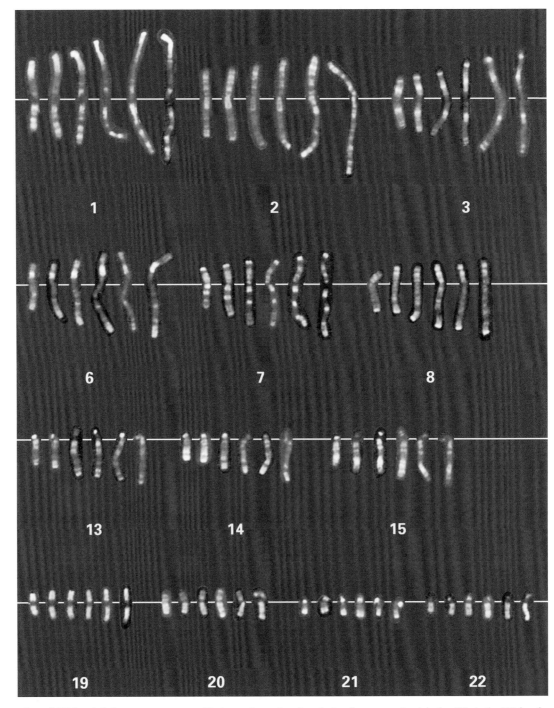

Fig. 6. (**b**) R-banded chromosomes arranged in increasing order of resolution from approximately the 400- to the 850-band levels. (Courtesy of Dr. E. Magenis.)

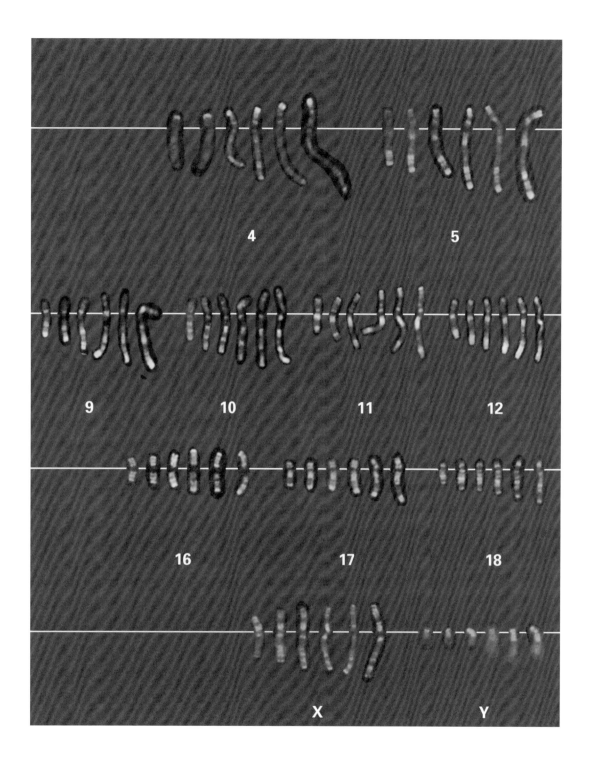

Normal Chromosomes

3 Symbols, Abbreviated Terms, and General Principles

All symbols and abbreviated terms used in the description of chromosomes and chromosome abnormalities are listed below. Section references are given within parentheses for terms that are defined in greater detail in the text. When more than one symbol or abbreviation is used together, a space is placed between the two (e.g., psu dic). When the symbol or abbreviation precedes the total number of chromosomes and no parenthesis is present, a space is placed between the symbol or abbreviation and the number of chromosomes (e.g., mos 47,XXX[25]/46,XX[5]). There is no space when a symbol or abbreviation immediately precedes or follows a parenthesis.

AI	First meiotic anaphase (12.1)
AII	Second meiotic anaphase (12.1)
ace	Acentric fragment (9.2.12, 10.2.1)
add	Additional material of unknown origin (9.2.1)
amp	Denotes an amplified signal (13.2.5, 13.3)
arr	Microarray (14.2)
arrow (→ or –>)	From – to, in detailed system (4.3.2.1)
b	Break (10.1.1, 10.2.1)
brackets, angle (< >)	Surround the ploidy level (8.1)
brackets, square ([ ])	Surround number of cells or genome build (4.1, 11.1.2, 14, 15, 16)
c	Constitutional anomaly (4.1, 8.3, 11.3)
cen	Centromere (2.3.2, 4.3.2.1)
cha	Chromoanasynthesis (14.2.7)
chi	Chimera (4.1)
chr	Chromosome (10.2)
cht	Chromatid (10.1, 14.2.8)
colon, single (:)	Break, in detailed form (4.3.2.1, 9.2)
colon, double (::)	Break and reunion, in detailed form (4.3.2.1, 9.2, 16.2)
comma (,)	Separates chromosome numbers, sex chromosomes, and chromosome abnormalities (4.1, 14.2); separates locus designations (13.2, 13.3.1)
con	Connected signals (13.3.5)
cp	Composite karyotype (11.1.5)
cth	Chromothripsis (14.2.7)
cx	Complex rearrangements (10.1.1, 14.2.7)
decimal point (.)	Denotes sub-bands (2.3.2)
del	Deletion (9.2.2, 16.3.1)
delins	Sequence change with nucleotides of the reference sequence replaced by other nucleotides (16.2, 16.3)
der	Derivative chromosome (4.4, 9.2.3, 9.2.17.2, 9.2.17.3, 16.3.2)
dia	Diakinesis (12.1)
dic	Dicentric (9.2.4)
dim	Diminished (13.2.6, 13.5)
dinh	Derived from chromosome abnormality of parental origin (4.1, 14.2.3)

dip	Diplotene (12.1)	
dis	Distal (12.1)	
dit	Dictyotene (12.1)	
dmat	Derived from chromosome abnormality of maternal origin (4.1, 14.2.3)	
dmin	Double minute (9.2.12, 10.2.1)	
dn	Designates a chromosome abnormality that has not been inherited (*de novo*) (4.1, 14.2.3)	
dpat	Derived from chromosome abnormality of paternal origin (4.1, 14.2.3)	
dup	Duplication (9.2.5, 16.3.3)	
e	Exchange (10.1.1, 10.2.1)	
enh	Enhanced (13.2.6, 13.5)	
equal sign (=)	Number of chiasmata (12.1)	
fem	Female (12.1)	
fib	Extended chromatin/DNA fiber (13.4)	
fis	Fission, at the centromere (9.2.6)	
fra	Fragile site (7.2, 9.2.7)	
g	Gap (10.1.1, 10.2.1)	
g.	Genome with reference to the genomic sequence (16)	
GRCh	Genome Reference Consortium human; human genome build or assembly (4.3, 14, 15, 16)	
greater than (>)	Greater than (9.2.12)	
h	Heterochromatin, constitutive (7.1.1)	
hmz	Homozygous, homozygosity; used when one or two copies of a genome are detected, but previous, known heterozygosity has been reduced to homozygosity through a variety of mechanisms, e.g., loss of heterozygosity (LOH) (14.2.6)	
hsr	Homogeneously staining region (9.2.8)	
htz	Heterozygous, heterozygosity (14.2.6)	
hyphen (-)	Hyphen, designation with a chromosome band at low resolution (9.2.2 and Figure 5)	
i	Isochromosome (9.2.11)	
idem	Denotes the stemline karyotype in a subclone (11.1.4)	
ider	Isoderivative chromosome (9.2.3)	
idic	Isodicentric chromosome (9.2.4, 9.2.11)	
inc	Incomplete karyotype (5.4)	
inh	Inherited (4.1, 14.2.3)	
ins	Insertion (9.2.9); insertion of nucleotides (16.2, 16.3.4)	
inv	Inversion (9.2.10); inverted in orientation relative to the reference sequence (16.2, 16.3.5)	
ish	*In situ* hybridization; when used without a prefix applies to metaphase or prometaphase chromosomes of dividing cells (13)	
lep	Leptotene (12.1)	
MI	First meiotic metaphase (12.1)	
MII	Second meiotic metaphase (12.1)	
mal	Male (12.1)	
mar	Marker chromosome (9.2.12)	
mat	Maternal origin (4.1, 14.2.3)	
med	Medial (12.1)	
min	Minute acentric fragment (10.2.1)	
minus sign (−)	Loss (4.1, 8.1); decrease in length (7.1.1); locus absent from a specific chromosome (13.2)	
mos	Mosaic (4.1, 14.2.5)	
multiplication sign (×)	Multiple copies of rearranged chromosomes (9.3); designates aberrant polyploidy clones in neoplasias (11.1.4); with number to indicate number of signals seen (13.2, 13.3); multiple copies of a chromosome or chromosomal region (14.2)	
neg	No presence of the rearrangement for which testing was conducted (15.4)	
neo	Neocentromere (9.2.13)	
nuc	Nuclear or interphase (13.3)	
oom	Oogonial metaphase (12.1)	
or	Alternative interpretation (5.3)	
p	Short arm of chromosome (2.3.2)	
PI	First meiotic prophase (12.1)	
pac	Pachytene (12.1)	

parentheses ( )	Surround structurally altered chromosomes and breakpoints (4.1); surround chromosome numbers, X, and Y in normal and abnormal results; surround coordinates (or nucleotide positions) in abnormal result (14, 15, 16)
pat	Paternal origin (4.1, 14.2.3)
pcc	Premature chromosome condensation (10.2.1)
pcd	Premature centromere division (10.2.1)
pcp	Partial chromosome paint (13.7)
period (.)	Separates various techniques (13.2, 14.2, 16.2, 16.3)
Ph	Philadelphia chromosome (9.2.3)
plus sign, single (+)	Additional normal or abnormal chromosomes (4.1, 8.1); increase in length (7.1.1); locus present on a specific chromosome (13.2)
plus sign, double (++)	Two hybridization signals or hybridization regions on a specific chromosome (13.2)
pos	Detection of a rearrangement for which testing was conducted (15.4)
prx	Proximal (12.1)
ps	Satellited short arm of chromosome (7.1.1, 7.1.2)
psu	Pseudo- (9.2.4)
pter	Terminal end of the short arm
pvz	Pulverization (10.2.1)
q	Long arm of chromosome (2.3.2)
qdp	Quadruplication (9.2.14)
qr	Quadriradial (10.1.1)
qs	Satellited long arm of chromosome (7.1.1, 7.1.2)
qter	Terminal end of the long arm
question mark (?)	Questionable identification of a chromosome or chromosome structure (5.1)
r	Ring chromosome; a defined structure with chromosome ends fused (9.2.15, 16.3.6)
rec	Recombinant chromosome (4.5, 9.2.3)
rev	Reverse (13.5)
rob	Robertsonian translocation (9.2.17.3)
Roman numerals I–IV	Indicate univalent, bivalent, trivalent, and quadrivalent structures (12.1)
rsa	Region-specific assay (15)
s	Satellite (7.1.1, 7.1.2)
sce	Sister chromatid exchange (10.1.1)
sdl	Sideline (11.1.4)
semicolon (;)	Separates altered chromosomes and breakpoints in structural rearrangements involving more than one chromosome (4.1, 4.3.1, 12.1); separates probes on different derivative chromosomes (13.2)
sep	Separated signals (13.3.5)
seq	Sequencing (16)
sl	Stemline (11.1.4)
slant line, single (/)	Separates clones (4.1, 11.1.1, 11.1.6, 11.3), or contiguous probes (13.2, 13.3)
slant line, double (//)	Separates chimeric clones (4.1, 13.3.4)
spm	Spermatogonial metaphase (12.1)
sseq	Shallow next-generation sequencing (14.2.8)
stk	Satellite stalk (7.1.1, 7.1.2)
subtel	Subtelomeric region (13.2.7)
sup	Additional (supernumerary) sequence not attached to other chromosomal material (16.2, 16.3.6)
t	Translocation (9.2.17, 16.3.7)
tas	Telomeric association (9.2.16)
ter	Terminal (end of chromosome) or telomere (4.3.2.1)
tilde (~)	Denotes intervals and boundaries of a chromosome segment or number of chromosomes, fragments, or markers (5.2); denotes a range of number of copies of a chromosomal region when the exact number cannot be determined (14.2)
tr	Triradial (10.1.1)
trc	Tricentric chromosome (9.2.18)
trp	Triplication (9.2.19)
underlining (single)	Used to distinguish homologous chromosomes (4.1, 9.2.3, 9.2.17.1)
underscore (_)	Used to indicate range of nucleotide positions (14, 15, 16)
upd	Uniparental disomy (8.4, 14.2.1)
var	Variant or variable region (2.4, 7.1)
wcp	Whole chromosome paint (13.2)
xma	Chiasma(ta) (12.1)
zyg	Zygotene (12.1)

The following general principles are applicable to multiple chapters and techniques as indicated:

General Principles

	Banded chromosomes	Oncology	FISH	Micro-array	Region-specific assays	Sequence-based
Sex chromosome abnormalities listed first	+	+	+	+	+	+
Breakpoint band designations from pter to qter	+	+	+	+	+	+
Different band resolutions may be used within the same karyotype string	+	+				
For each chromosome, numerical abnormalities are listed before structural changes	+	+				
Multiple structural changes presented in alphabetical order	+	+				
Ring chromosomes presented before marker chromosomes	+	+				
Number of cells in each cell line shown in square brackets for mosaic	+			+		
Number of cells shown in square brackets for oncology (both single or multiple clones)		+	+			
Related clones listed in order of increasing complexity, irrespective of size		+				
Largest cell line/unrelated clone is listed first; normal cells are always listed last	+	+	+			
Express abnormalities relative to the appropriate ploidy level	+	+	+	+	+	+
Separate results of different techniques with periods (.)	+	+	+	+	+	+
Genome build required when nucleotides designated				+	+	+
Nucleotide numbers given either with or without commas to indicate thousands and millions for structural anomalies				+	+	+
Nucleotide span separated by an underscore				+	+	+
Proportion of DNA shown in square brackets				+	+	+

+, applicable.

4 Karyotype Designation

4.1 General Principles

In the description of a karyotype the first item to be recorded is the total number of chromosomes, including the sex chromosomes, followed by a **comma** (,). The sex chromosome constitution is given next. The autosomes are specified only when an abnormality is present. Thus, the normal human karyotype is designated as follows:

46,XX Normal female
46,XY Normal male
46,X? Where sex is not to be disclosed

- In the description of chromosome abnormalities, sex chromosome aberrations are presented first, followed by abnormalities of the autosomes listed in numerical order irrespective of aberration type. Each abnormality is separated by a comma. Details regarding the order of chromosome abnormalities are presented in Chapter 6.
- Letter designations are used to specify rearranged (i.e., structurally altered) chromosomes. All symbols and abbreviations used to designate chromosome abnormalities are listed in Chapter 3.
- In single chromosome rearrangements, the chromosome involved in the change is specified within **parentheses** ( ) immediately following the symbol identifying the type of rearrangement, e.g., inv(2), del(4), r(18). If two or more chromosomes have been altered, a **semicolon** (;) is used to separate their designations. If one of the rearranged chromosomes is a sex chromosome, then it is listed first; otherwise, the chromosome having the lowest number is always specified first, e.g., t(X;3) or t(2;5). An exception to this rule involves certain three-break rearrangements in which part of one chromosome is inserted at a point of breakage in another chromosome. In this event, the receptor chromosome is specified first, regardless of whether it is a sex chromosome or an autosome with a number higher or lower than that of the donor chromosome, e.g., ins(5;2). For details, see Section 9.2.9.
- For balanced translocations involving three separate chromosomes, with one breakpoint in each chromosome, the rule is still followed that the sex chromosome or autosome with the lowest number is specified first. The chromosome listed next is the one that receives a segment from the first chromosome, and the chromosome specified last is the one that donates a segment to the first listed chromosome. The same rule is followed in four-break and more complex balanced translocations (see also Section 9.2.17.1).

- In order to distinguish homologous chromosomes, one of the numerals may be underlined (**single underlining**). The derivative chromosomes produced by reciprocal translocations should be described using the conventions outlined in Section 9.2.3.
- A **plus** (+) or **minus** (−) sign is placed before a chromosome or an abnormality designation to indicate additional or missing, normal or abnormal chromosomes, e.g., +21, −7, +der(2); for details, see Section 8.1.
- The + or − sign placed after a chromosome arm symbol (p or q) may be used in text to indicate an increase or decrease in the length of a chromosome arm (e.g., 4p+, 5q−) but should not be used in the description of karyotypes. See also Sections 9.2.1 and 9.2.2. Variations in length of heterochromatic segments, satellites, and satellite stalks are distinguished from increases or decreases in arm length as a result of other structural alterations by placing a plus or minus sign after the appropriate symbol for these normal variable chromosome features (see Section 7.1). The use of + and − signs in the description of results obtained by *in situ* hybridization is described in Chapter 13.
- When normal chromosomes are replaced by structurally altered chromosomes, the normal ones should not be recorded as missing (see Section 9.1). In the description of karyotypes containing dicentric chromosomes or derivative chromosomes resulting from whole-arm translocations, the abnormal chromosomes by convention replace both normal chromosomes involved in the formation of the dicentric chromosome or derivative chromosome. Thus, in these situations the two missing chromosomes are not specified (see Sections 9.2.4 and 9.2.17.2).
- The **multiplication sign** (×) can be used to describe multiple copies of a rearranged chromosome but should not be used to denote multiple copies of normal chromosomes (see Section 9.3).
- Uncertainty in chromosome or band designation may be indicated by a **question mark** (?) or a **tilde** (~). The term **or** is used to indicate alternative interpretations of an aberration. For details, see Chapter 5.
- The karyotype designations of different clones or cell lines are separated by a **slant line** (/).
- **Square brackets** [ ], placed after the karyotype description, are used to designate the absolute number of cells in each cell line or clone (see Sections 8.1, 8.2, 8.3, and 11.1.2).
- In order to distinguish between a *mosaic* (cell lines originating from the same zygote) and a *chimera* (cell lines originating from different zygotes) in constitutional cases, the symbol **mos** or **chi**, respectively, preceding the karyotype designations, may be used; for example, mos 45,X[10]/46,XX[10] and chi 46,XX[10]/46,XY[10]. In most instances the abbreviations will be needed only for the initial description in any report; subsequently, the simple karyotype designation may be used. A space should follow mos or chi.
- All abbreviations that precede a number will have a space that follows.
- A normal diploid cell line, when present, is always listed last, e.g., mos 47,XY,+21[10]/46,XY[10]; mos 47,XXY[10]/46,XY[10].
- If there are several abnormal cell lines, they are presented according to their size; the largest first, then the second largest, and so on, e.g., mos 45,X[15]/47,XXX[10]/46,XX[23]. Likewise, the largest cell line in chimeras is presented first, e.g., chi 46,XX[25]/46,XY[10]. When equivalent numbers of cells are found in two cell lines, one of which has a numerical abnormality and the other of which has a structural abnormality, the numerical is listed first, e.g.,45,X[25]/46,X,i(X)(q10)[25]. When both cell lines have numerical abnormalities they are listed according to the altered autosome number, e.g., 47,XX,+8[25]/47,XX,+21[25]; a cell line with a sex chromosome abnormality always comes first, e.g., 47,XXX[25]/47,XX,+21[25]. For order of clone presentation in neoplasia, see Sections 11.1.2, 11.1.4, and 11.1.6.

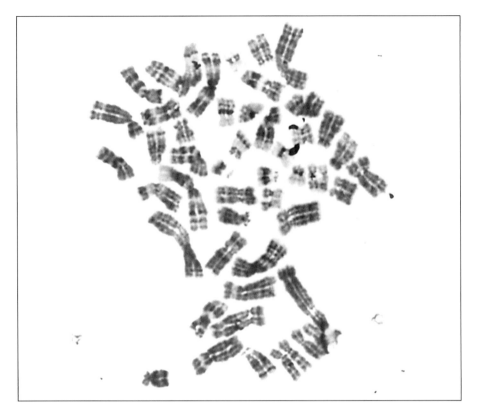

Fig. 7. Metaphase chromosomes in a cell which has undergone endoreduplication. (Courtesy of Dr. N. Mandahl.)

- In chimerism secondary to stem cell transplant, the recipient cell clones are listed first, followed by the donor cell line(s). The recipient and donor cell line(s) are separated by a **double slant line** (//) as shown in the following examples.

  46,XY[3]//46,XX[17]
  Three cells from the male recipient were identified along with 17 cells from the female donor.

  46,XY,t(9;22)(q34;q11.2)[4]//46,XX[16]
  Four male recipient cells showing a 9;22 translocation were identified along with 16 female donor cells.

  //46,XX[20]
  All 20 cells were identified as derived from the female donor.

  46,XY[20]//
  All 20 cells were identified as derived from the male recipient.

- A haploid or polyploid karyotype will be evident from the chromosome number and from the further designations, e.g., 69,XXY. All chromosome changes should be expressed in relation to the appropriate ploidy level (see Sections 8.1 and 9.1), e.g., 70,XXY,+21.
- Endoreduplication is the replication of the chromosomes without chromatid separation or cytokinesis (Fig. 7). Technologies such a SNP microarray may be helpful in differentiating between endoreduplication and other mechanisms (e.g., aborted mitosis and cell fusion) which may produce a similar copy number change.

- When it is known that a particular chromosome involved in an aberration has been inherited, the term **inh** may be used, e.g., 46,XX,t(5;6)(q34;q23)inh. When it is known that the aberration is inherited from the mother or the father, the most complete information is evident by the use of the abbreviation **mat** or **pat**, respectively, immediately following the designation of the abnormality, e.g., 46,XX,t(5;6)(q34;q23)mat,inv(14)(q12q31)pat. However, if only part of an aberration (e.g., one derivative chromosome from a balanced parental translocation) has been inherited, the abbreviation **dmat**, **dpat** or **dinh** is used to distinguish it from the complement in the parent. If it is known that the parents' chromosomes are normal with respect to the abnormality, the abnormality may be designated *de novo* (**dn**), e.g., 46,XY,t(5;6)(q34;q23)mat,inv(14)(q12q31)dn. When dn follows another abbreviation, a space is inserted, e.g., 47,XY,+mar dn[14]/46,XY[16].
- The same rules for designating chromosome aberrations are followed in the description of constitutional and acquired chromosome aberrations. Terms and recommendations related to abnormalities seen in neoplasia are described in Chapter 11.
- When an acquired chromosome abnormality is found in an individual with a constitutional chromosome anomaly, the latter is indicated by the letter **c** immediately after the constitutional abnormality designation (see Sections 8.3 and 11.3).
- In the interest of clarity, complex rearrangements necessitating lengthy descriptions should be written out in full the first time they are used in a report. An abbreviated version may be used subsequently, provided it is clearly defined immediately after the complete notation.
- Nomenclature guidelines for meiotic chromosomes are presented in Chapter 12.

4.2 Specification of Breakpoints

The location of any given breakpoint is specified by the band in which that break has occurred. Since it is not possible at present to define band interfaces accurately, a break suspected to be at an interface between two bands is assigned arbitrarily to the higher of the two band numbers, i.e., the number of the band more distal to the centromere.

A given break may sometimes appear to be located in either of two consecutive bands. A similar situation may occur when breaks at or near an interface between two bands are studied with two or more techniques. In this event, the break can be specified by both band numbers separated by the term **or**, e.g., 1q23 or q24, indicating a break in either band 1q23 or band 1q24 (see also Section 5.3). If a break can be localized to a region but not to a particular band, only the region number may be specified, e.g., 1p1. Uncertainty about breakpoint localization may also be indicated by a question mark, e.g., 1p1? (see Section 5.1). If the breakpoint can be assigned only to two adjacent regions, both suspected regions should be indicated, e.g., 1q2 or q3. For the use of the tilde to express uncertainty, see Section 5.2. Breakpoints within the same rearrangement or karyotype string can be at different levels of resolution reflecting the precision of the karyogram; however, if an overall banding resolution is stated it should be done so in accordance with national regulatory agencies as applicable.

When an extra copy of a rearranged chromosome is present, the breakpoints do not need to be repeated; the breakpoints are specified only at the first time they appear in the karyotype, e.g., 48,XX,+1,+der(1)t(1;16)(p13;q13),t(1;16).

4.3 Designating Structural Chromosome Aberrations by Breakpoints and Band Composition

Two systems for designating structural abnormalities exist. One is a **short form** in which the nature of the rearrangement and the breakpoint(s) are identified by the bands or regions in which the breaks occur. Because of the conventions built into this form, the band composition of the abnormal chromosomes can readily be inferred from the information provided in the symbolic description. For very complex abnormalities, especially in tumor cells, the short form may be inadequate or ambiguous, but it will always provide information on all bands involved in the generation of an abnormal chromosome. The other is a **detailed form** which, besides identifying the type of rearrangement, defines each abnormal chromosome in terms of its band composition. The notation used to identify the rearrangement and the method of specifying the breakpoints are common to both forms (see Sections 4.3.1 and 4.3.2).

The genome build or assembly must be specified when designating an aberration by nucleotide position defined by microarray, region-specific assay or sequencing (see Chapters 14, 15, and 16).

4.3.1 Short Form for Designating Structural Chromosome Aberrations

In the short form, structurally altered chromosomes are defined only by their breakpoints. The breakpoints are specified within parentheses immediately following the designation of the type of rearrangement and the chromosome(s) involved. The breakpoints are identified by band designations and are listed in the same order as the chromosomes involved. No semicolon is used between breakpoints in single chromosome rearrangements.

4.3.1.1 Two-Break Rearrangements

Whether the two breaks occur within the same arm or in both arms of a single chromosome in a two-break rearrangement, the breakpoints are specified from pter to qter.

46,XX,inv(2)(p21q31)
46,XX,inv(2)(p23p13)
46,XX,inv(2)(q11.2q32)

When two chromosomes are involved, the chromosome having the lowest number is always listed first; however, if one of the rearranged chromosomes is a sex chromosome this is listed first.

46,XY,t(12;16)(q13;p11.1)
46,X,t(X;18)(p11.1;q11.1)

4.3.1.2 Three-Break Rearrangements

An exception to the rule that sex chromosomes and autosomes with the lowest number are specified first involves three-break rearrangements in which part of one chromosome is inserted into another chromosome. In that event, the donor chromosome is listed last, even if it is a sex chromosome or an autosome with a lower number than that of the receptor chromosome.

46,X,ins(5;X)(p14;q21q25)
46,XY,ins(5;2)(p14;q22q32)

When an insertion within a single chromosome occurs, the breakpoint at which the chromosome segment is inserted is always specified first. The remaining breakpoints are specified in the same way as in a two-break rearrangement, i.e., the breakpoints of the inserted segment are specified with the one closer to the pter of the recipient chromosome listed first.

46,XX,ins(2)(q13p13p23)
   Insertion of the short-arm segment between bands 2p23 and 2p13 into the long arm at band 2q13 with 2p13 being closer to pter than 2p23.

46,XX,ins(2)(q13p23p13)
   Insertion of the short-arm segment between bands 2p23 and 2p13 into the long arm at band 2q13 with 2p23 being closer to pter than 2p13.

For translocations involving three chromosomes, with one breakpoint in each, the rule is still followed that the sex chromosome or autosome with the lowest number is given first. The chromosome listed next is the one that receives a segment from the first chromosome, and the chromosome specified last is the one that donates a segment to the first chromosome listed.

46,XX,t(9;22;17)(q34;q11.2;q22)
   The segment of chromosome 9 distal to 9q34 has been translocated onto chromosome 22 at band 22q11.2, the segment of chromosome 22 distal to 22q11.2 has been translocated onto chromosome 17 at 17q22, and the segment of chromosome 17 distal to 17q22 has been translocated onto chromosome 9 at 9q34.

46,Y,t(X;15;18)(p11.1;p11.1;q11.1)
   The segment of the X chromosome distal to Xp11.1 has been translocated onto chromosome 15 at band 15p11.1, the segment of chromosome 15 distal to 15p11.1 has been translocated onto chromosome 18 at 18q11.1, and the segment of chromosome 18 distal to 18q11.1 has been translocated to Xp11.1.

4.3.1.3 Four-Break and More Complex Rearrangements

Whenever applicable, the guidelines for three-break rearrangements should be used.

46,XX,t(3;9;22;21)(p13;q34;q11.2;q21)
   The segment of chromosome 3 distal to 3p13 has been translocated onto chromosome 9 at 9q34, the segment of chromosome 9 distal to 9q34 has been translocated onto chromosome 22 at 22q11.2, the segment of chromosome 22 distal to 22q11.2 has been translocated onto chromosome 21 at 21q21, and the segment of chromosome 21 distal to 21q21 has been translocated onto chromosome 3 at 3p13.

46,XY,t(5;6)(q13q23;q15q23)
   Reciprocal translocation of two interstitial segments. The segments between bands 5q13 and 5q23 of chromosome 5 and between 6q15 and 6q23 of chromosome 6 have been exchanged.

Unbalanced rearrangements will lead to at least one derivative chromosome and in these situations the use of the symbol **der** to describe the derivative chromosome(s) is recommended (see Sections 4.4 and 9.2.3). It will usually not be possible to adequately describe all complex rearrangements with the short form. The detailed form can always be used to describe any abnormality, however complex. Still, it may be necessary to illustrate the rearrangement and/or describe it in words to ensure complete clarity.

4.3.2 Detailed Form for Designating Structural Chromosome Aberrations

Structurally altered chromosomes are defined by their band composition. The conventions used in the short form are retained in the detailed form, except that an abbreviated description of the band composition of the rearranged chromosome(s) is specified within the last parentheses, instead of only the breakpoints. It is acceptable to combine the short form (4.3.1) and the detailed form for designating complex karyotypes, especially to describe acquired chromosomal abnormalities.

4.3.2.1 Additional Symbols

A **single colon** (:) is used to indicate a chromosome *break* and a **double colon** (::) to indicate *break and reunion*. In order to avoid an unwieldy description, an **arrow** (→ or –>), meaning *from – to*, is employed. The end of a chromosome arm may be designated either by its band designation or by the symbol **ter** *(terminal)*, preceded by the arm designation, i.e., **pter** indicates the end of the short arm and **qter** the end of the long arm. When it is necessary to indicate the *centromere*, the abbreviation **cen** should be used.

4.3.2.2 Designating the Band Composition of a Chromosome

The description starts at the end of the short arm and proceeds to the end of the long arm, with the bands being identified in the order in which they occur in the rearranged chromosome. If the rearrangement is confined to a single chromosome, the chromosome number is not repeated in the band description. If more than one chromosome is involved, however, the bands and chromatid ends are identified with the appropriate chromosome numbers. The aberrations should be listed according to the breakpoints of the derivative chromosome from pter to qter and should not be separated by a comma.

If, owing to a rearrangement, no short-arm segment is present at the end of either arm, the description of the structurally rearranged chromosome starts at the end of the long-arm segment with the lowest chromosome number. However, if a portion of the proximal short arm is present, the description begins with the material on the end of that chromosome arm even if the recipient segment is from a long arm or from a chromosome with a higher or lower chromosome number. For use of the detailed form, see examples in Section 9.2.3.

4.4 Derivative Chromosomes

A *derivative chromosome* is a structurally rearranged chromosome generated by (1) more than one rearrangement within a single chromosome, e.g., an inversion and a deletion of the same chromosome, or deletions in both arms of a single chromosome, or (2) rearrangements involving two or more chromosomes, e.g., the unbalanced product(s) of a translocation. An abnormal chromosome in which no part can be identified is referred to as a marker chromosome (see Section 9.2.12).

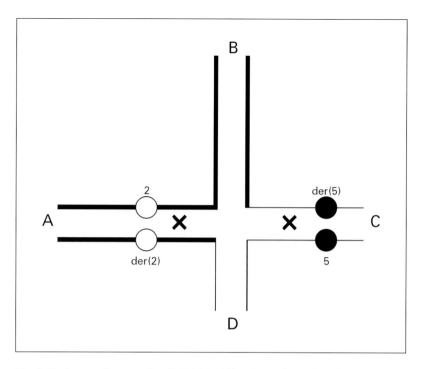

Fig. 8. Pachytene diagram of a t(2;5)(q21;q31) reciprocal translocation heterozygote used to specify the disjunctional possibilities and derivative chromosome combinations given in Table 4. Letters A, B, C, and D designate chromosome ends (telomeres). For the sake of simplicity only those two of the four chromatids that are involved in crossing-over (see Table 4) are indicated. Crosses mark the positions of crossing-over.

Derivative chromosomes are designated **der**. The term always refers to the chromosome(s) that has an intact centromere or neocentromere (Section 9.2.13). The derivative chromosome is specified in parentheses, followed by all aberrations involved in the generation of the derivative chromosome. The aberrations should be listed according to the breakpoints of the derivative chromosome from pter to qter and should not be separated by a comma. For example, der(1)t(1;3)(p32;q21)t(1;11)(q25;q13) specifies a derivative chromosome 1 generated by two translocations, one involving the short arm with a breakpoint in 1p32 and the other involving the long arm with a breakpoint in 1q25.

Various derivative chromosomes and their designations are presented in Section 9.2.3. As an illustration of the way derivative chromosomes can be written, a balanced reciprocal translocation between chromosomes 2 and 5, 46,XX,t(2;5)(q21;q31), has been assumed and is represented by the pachytene diagram in Fig. 8. The derivative chromosomes from such a translocation would be designated der(2) and der(5). Table 4 gives the possible unbalanced gametes resulting from adjacent-1 and adjacent-2 disjunctions and also from four of the 12 possible 3:1 disjunctions, together with the recommended designations of the karyotypes resulting from syngamy between each unbalanced gametic type and a normal gamete. The full karyotype designation needs be written only once in any given publication and then can be abbreviated (i.e., breakpoints not included). A suggested abbreviation for the first designated karyotype in Table 4, for example, would be 46,XX,der(5)dmat.

Table 4. Possible unbalanced gametes derived from segregation of a balanced reciprocal translocation of maternal origin. The pachytene configuration is given in Fig. 8.

Segregation pattern	Schematic segregants	Chromosomal complement of gametes	Karyotype of potential female zygotes
Adjacent-1	AB CB	2, der(5)	46,XX,der(5)t(2;5)(q21;q31)dmat
	AD CD	der(2), 5	46,XX,der(2)t(2;5)(q21;q31)dmat
Adjacent-2[a]	AB AD	2, der(2)	46,XX,+der(2)t(2;5)(q21;q31)dmat,−5
	CD CB	5, der(5)	46,XX,−2,+der(5)t(2;5)(q21;q31)dmat
	AB AB	2, 2	46,XX,+2,−5
	AD AD	der(2), der(2)	46,XX,der(2)t(2;5)(q21;q31)dmat,+der(2)t(2;5),−5
	CB CB	der(5), der(5)	46,XX,−2,der(5)t(2;5)(q21;q31)dmat,+der(5)t(2;5)
	CD CD	5, 5	46,XX,−2,+5
3:1[b]	AB CD CB	2, 5, der(5)	47,XX,+der(5)t(2;5)(q21;q31)dmat
	AD	der(2)	45,XX,der(2)t(2;5)(q21;q31)dmat,−5
	AD CD CB	der(2), 5, der(5)	47,XX,t(2;5)(q21;q31)mat,+5
	AB	2	45,XX,−5
	AB AD CD	2, der(2), 5	47,XX,+der(2)t(2;5)(q21;q31)dmat
	CB	der(5)	45,XX,−2,der(5)t(2;5)(q21;q31)dmat
	AB AD CB	2, der(2), der(5)	47,XX,+2,t(2;5)(q21;q31)mat
	CD	5	45,XX,−2

[a] Adjacent-2 disjunction minimally results in the first two unbalanced gametic types shown (AB AD, CD CB). Crossing-over in the interstitial segments between centromeres and points of exchange is necessary for the origin of the remaining four types.

[b] A further eight segregants can occur if there is crossing-over in the interstitial segments, making a total of 12 types of gametes with three chromosomes derived from the translocation quadrivalent.

4.5 Recombinant Chromosomes

A *recombinant chromosome* is a structurally rearranged chromosome with a new segmental composition resulting from *meiotic* crossing-over between a displaced segment and its normally located counterpart in certain types of structural heterozygotes.

Whereas derivative chromosomes are products of the original rearrangement and segregate at meiosis without further change, recombinant chromosomes arise *de novo* during gametogenesis in appropriate structural heterozygotes as predictable consequences of crossing-over in a displaced segment.

Recombinant chromosomes are designated by the symbol **rec**. The recombinant chromosome is specified in parentheses immediately following the symbol. The chromosome designation used is that which indicates the origin of the centromere of the particular recombinant chromosome.

Recombinant chromosomes are most likely to originate from crossing-over in inversion or insertion heterozygotes. To exemplify the method of designating these chromosomes, a maternal pericentric inversion of chromosome 2, 46,XX,inv(2)(p21q31), is shown diagrammatically in Fig. 9. In this case, crossing-over results in a duplication (dup) of 2p in one recombinant chromosome and of 2q in the other. The respective karyotype could be recorded as 46,XX,rec(2)dup(2p)inv(2)(p21q31)dmat and 46,XX,rec(2)dup(2q)inv(2)(p21q31)dmat, specifying, in the first example, a duplication from 2pter to 2p21 and a deletion from 2q31 to 2qter and, in the second example, a duplication from 2q31 to 2qter and a deletion from 2pter to 2p21. Note that, in analogy with the nomenclature for derivative chromosomes, the

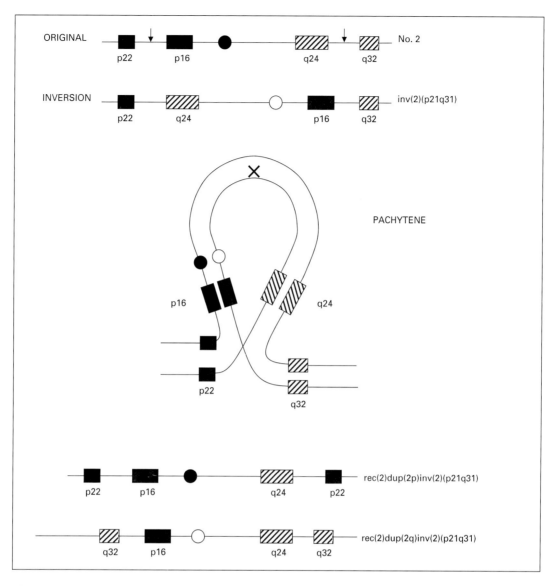

Fig. 9. Diagram of an inv(2)(p21q31)mat pericentric inversion heterozygote. Bands delimiting the breakpoints (arrows on original) are shown as black boxes on the short arm and as hatched boxes on the long arm. In the pachytene diagram, the cross indicates crossing-over within the inversion loop. For the sake of simplicity only those two of the four chromatids that are involved in crossing-over and give rise to the recombinant chromosomes are indicated.

aberrations following the designation rec are not separated by a comma. The symbol rec should only be used when a parental inversion or insertion has been identified. If this is not known, an apparent recombinant chromosome should be written as a derivative. For example, 46,XX,rec(2)dup(2p)inv(2)(p21q31)dmat designates a recombinant from a known maternal inversion. 46,XX,der(2)(pter→q31::p21→pter) designates a derivative chromosome with duplication of pter→p21 and deletion of q31→qter. The net imbalance is the same in the two examples, with the first derived from a known inversion carrier.

5 Uncertainty in Chromosome or Band Designation

5.1 Questionable Identification

A **question mark (?)** indicates questionable identification of a chromosome or chromosome structure. It is placed either **before the uncertain item**, or it may replace a chromosome, region, or band designation (see also examples in Section 9.2.3).

45,XX,–?21
   A missing chromosome, probably No. 21.

47,XX,+?8
   An additional chromosome, probably No. 8.

46,XX,del(1)(q2?)
   The break in the long arm of chromosome 1 is in region 1q2, but it has not been possible to determine the band within that region.

46,XY,del(1)(q2?3)
   The break in the long arm of chromosome 1 is in region 1q2, probably in band 1q23, but this is uncertain.

46,XX,del(1)(q?2)
   The break is in the long arm of chromosome 1, probably in region 1q2.

46,XY,del(1)(q?23)
   It is uncertain whether the break in the long arm of chromosome 1 is in region 1q2. If so, the break is in band 1q23.

46,XX,del(1)(q?)
   The break is in the long arm of chromosome 1, but neither the region nor the band can be identified. This aberration is often described as 1q–, a notation that may be useful in text but should not be used in karyotype nomenclature.

46,XY,?del(1)(q23)
   A possible deletion in chromosome 1, band 1q23, but all items, including the deletion, are uncertain.

46,XX,der(1)?t(1;3)(p22;q13)
: The der(1) has probably resulted from a t(1;3). If so, the breaks are in bands 1p22 and 3q13.

46,XY,der(5)ins(5;?)(q32;?)
: A derivative chromosome from the insertion of unidentified chromosomal material into the long arm of chromosome 5 at band q32.

5.2 Uncertain Breakpoint Localization or Chromosome Number

A **tilde** (~) is used to denote intervals and to express uncertainty about breakpoint localizations in that it indicates the boundaries of a chromosome segment in which the breaks may have occurred.

46,XX,del(1)(q21~24)
: A terminal deletion of the long arm of chromosome 1 with a breakpoint within the segment 1q21–q24, i.e., the breakpoint may be in band 1q21, 1q22, 1q23 or 1q24.

46,XY,dup(1)(q22~24q44)
: A duplication in the long arm of chromosome 1; the proximal breakpoint is in band 1q22, 1q23 or 1q24.

46,XX,t(3;12)(q27~29;q13~15)
: Both breakpoints in this translocation are uncertain; in chromosome 3 the breakpoint may be in bands 3q27, 3q28 or 3q29 and in chromosome 12 in bands 12q13, 12q14 or 12q15.

43~47,XX,…
: The chromosome number is within the interval 43–47.

5.3 Alternative Interpretation

The symbol **or** is used to indicate alternative interpretations of an aberration. Note that there should be a space before and after the symbol.

46,XX,add(19)(p13 or q13)
: Additional material of unknown origin attached to either 19p13 or 19q13 (see Section 9.2.1).

46,XY,del(8)(q21.1) or i(8)(p10)
: A deletion of the long arm of a chromosome 8 with a breakpoint in 8q21.1 or an isochromosome for the short arm of chromosome 8.

46,XX,t(12;14)(q15;q24) or t(12;14)(q13;q22)
: The two alternative interpretations of the t(12;14) give rise to identical-looking derivative chromosomes. This is in principle a different situation than t(12;14)(q13~15;q22~24), which means that the breakpoint localizations in the t(12;14) are less certain and a variety of combinations are possible.

46,XY,der(1)t(1;10)(q44;q22) or dup(1)(q32q44)
: A rearranged chromosome 1 that may have originated either from a translocation to 1q44 of the distal segment of the long arm of chromosome 10 with a breakpoint in 10q22, or from a duplication of the segment from 1q32 to 1q44. Further testing is required to differentiate between the two possibilities.

5.4 Incomplete Karyotype

The symbol **inc** denotes that the karyotype presented is incomplete, usually because of poor chromosome quality. The karyotype is thus likely to contain unidentified structural or numerical changes in addition to the abnormalities listed. The symbol **inc** is placed at the end of the nomenclature string, after the description of identifiable abnormalities.

46,XX,del(1)(q21),inc[4]
    It has only been possible to identify a clonally occurring deletion of the long arm of chromosome 1, but analysis is incomplete. Without the symbol inc, the del(1)(q21) would be the sole anomaly present in this tumor.

53~57,XY,+1,+3,+6,t(9;22)(q34;q11.2),+21,+3mar,inc[cp10]
    This abnormal karyotype has, in addition to the abnormalities presented that include three marker chromosomes, other changes that could not be identified. cp indicates a composite karyotype from 10 cells (see Section 11.1.5).

Every attempt should be made to present karyotypes in which each abnormality has been identified. The use of **inc** should be restricted to exceptional situations.

6 Order of Chromosome Abnormalities in the Karyotype

Sex chromosome aberrations are specified first (X chromosome abnormalities are presented before those involving Y), followed by abnormalities of the autosomes listed in numerical order irrespective of aberration type. For each chromosome, numerical abnormalities are listed before structural changes. Multiple structural changes of homologous chromosomes are presented in alphabetical order according to the abbreviated term of the abnormality. For order of clone presentation, see Sections 4.1, 11.1.4 and 11.1.6.

50,X,+X,−Y,+10,+14,+17,+21[5]/46,XY[15]
    The numerical abnormality of the X is listed before that of the Y.

47,X,t(X;13)(q27;q12),inv(10)(p13q22),+21
    The sex chromosome abnormality is presented first, followed by the autosomal abnormalities in chromosome number order, irrespective of whether the aberrations are numerical or structural.

47,Y,t(X;13)(q27;q12),inv(10)(p13q22),+21
    The same karyotype as in the previous example in a male.

46,t(X;18)(p11.1;q11.2),t(Y;1)(q11.2;p13)
    The abnormality involving the X chromosome is listed before that of the Y chromosome.

48,X,t(Y;12)(q11.2;p12),del(6)(q11),+8,t(9;22)(q34;q11.2),+17,−21,+22
    The translocation involving the Y chromosome is presented first, followed by all autosomal abnormalities in strict chromosome number order.

49,X,inv(X)(p21q26),+3,inv(3)(q21q26.2),+7,+10,−20,del(20)(q11.2),+21
    The inversion of the X chromosome is listed first. The extra chromosome 3 is presented before the inversion of chromosome 3 and the monosomy 20 before the deletion of chromosome 20. Note that the karyotype can also be written as: 49,X,inv(X)(p21q26),+inv(3)(q21q26.2),+7,+10,−20,del(20)(q11.2),+21 and is preferred as it gives the shortest karyotype string.

50,XX,+1,+del(1)(p13),+dup(1)(q21q32),+inv(1)(p31q41),+8,r(10)(p12q25),−21
    There are four abnormalities involving different copies of chromosome 1. The numerical change is presented first, followed by the structural aberrations listed in alphabetical order: del, dup, inv.

46,XX,der(8)ins(8;?)(p23;?)del(8)(q22)
    There are two abnormalities involving one chromosome 8. The chromosome 8 is described as a derivative with the structural aberrations listed from the distal p arm to the distal q arm, rather than in alphabetical order, because the insertion and deletion are present on the same derivative chromosome.

Unidentified ring chromosomes (r), marker chromosomes (mar), and double minute chromosomes (dmin) are listed last, in that order.

52,XX,…,+r,+mar,12~20dmin

Derivative chromosomes whose centromere is unknown (see Section 9.2.3) should be placed after all identified abnormalities but before unidentified ring chromosomes, marker chromosomes, and double minute chromosomes.

52,XX,…,+der(?)t(?;6)(?;q16),+r,+mar,5~9dmin

7 Normal Variable Chromosome Features

7.1 Variation in Heterochromatic Segments, Satellite Stalks, and Satellites

Variation refers to the differences in size or staining of chromosomal segments in the population (see Wyandt and Tonk, 2008, and Table 1). The following sections outline a means to describe these variations; however, to avoid misinterpretation, it is strongly recommended that these variants not be included in ISCN nomenclature descriptions. Rather, they should be reserved for report text descriptions as required. The variation may be useful to distinguish between two or more distinct cell lines or clones, e.g., chimerism or transplant.

7.1.1 Variation in Length

Variation in length of *heterochromatic segments* (**h**), *stalks* (**stk**) or *satellites* (**s**) should be distinguished from increases or decreases in arm length as a result of other structural alterations by placing a **plus** (+) or **minus** (–) sign after the symbols **h**, **stk** or **s** following the appropriate chromosome and arm designation.

16qh+	Increase in length of the heterochromatin on the long arm of chromosome 16.
Yqh–	Decrease in length of the heterochromatin on the long arm of the Y chromosome.
21ps+	Increase in length of the satellite on the short arm of chromosome 21.
22pstk+	Increase in length of the stalk on the short arm of chromosome 22.
13cenh+pat	Increase in length of the centromeric heterochromatin of the chromosome 13 inherited from the father.
1qh–,13cenh+,22ps+	Decrease in length of the heterochromatin on the long arm of chromosome 1, increase in length of the centromeric heterochromatin on chromosome 13, and large satellites on chromosome 22.
15cenh+mat,15ps+pat	Increase in length of the centromeric heterochromatin on the chromosome 15 inherited from the mother and large satellites on the chromosome 15 inherited from the father.
14cenh+pstk+ps+	Increase in length of the centromeric heterochromatin, the stalk, and the size of satellites on the same chromosome 14.

7.1.2 Variation in Number and Position

The same nomenclature symbols as described above are used to describe variation in position of heterochromatic segments, satellite stalks, and satellites.

22pvar	Variable presentation of the short arm of chromosome 22.
17ps	Satellites on the short arm of chromosome 17.
Yqs	Satellites on the long arm of the Y chromosome.
9phqh	Heterochromatin in both the short and the long arms of chromosome 9.
9ph	Heterochromatin only in the short arm of chromosome 9.
1q41h	Heterochromatic segment in chromosome 1 at band 1q41.

Duplicated chromosome structures are indicated by repeating the appropriate designation:

21pss	Double satellites on the short arm of chromosome 21.
14pstkstk	Double stalks on the short arm of chromosome 14.

In contrast, the common population inversion variants (see Table 1) are specified by their euchromatic breakpoints.

inv(9)(p12q13)	Pericentric inversion on chromosome 9.
inv(2)(p11.2q13)	Pericentric inversion on chromosome 2.

7.2 Fragile Sites

Fragile sites (**fra**) associated with a specific disease or phenotype are referred to in Section 9.2.7.

Fragile sites associated with specific chromosome bands can occur as normal variants with no phenotypic consequences. These fragile sites are inherited in a co-dominant Mendelian fashion and may result in chromosome abnormalities such as deletions, multiradial figures, and acentric fragments. While there may be several different types of fragile sites inducible by culturing cells in media containing different components, all these will be covered by a single nomenclature.

fra(10)(q25.2)	A fragile site on chromosome 10 in 10q25.2.
fra(10)(q22.1),fra(10)(q25.2)	Two fragile sites on the same chromosome 10.
fra(10)(q22.1),fra(<u>10</u>)(q25.2)	Two fragile sites on different homologous chromosomes.
fra(10)(q25.2),fra(16)(q22.1)	Two fragile sites on different chromosomes.

8 Numerical Chromosome Abnormalities

8.1 General Principles

A **plus** (+) or **minus** (–) sign is placed before a chromosome to indicate gain or loss of that particular chromosome. The only exception to this rule is the convention to designate constitutional numerical sex chromosome abnormalities by listing all sex chromosomes after the chromosome number, see examples below.

All numerical changes are expressed in relation to the appropriate ploidy level (see Section 11.2), i.e., in near-haploid cells (chromosome numbers up to 34) in relation to 23, in near-diploid cells (chromosome numbers 35–57) in relation to 46, in near-triploid cells (chromosome numbers 58–80) in relation to 69, in near-tetraploid cells (chromosome numbers 81–103) in relation to 92, and so on.

26,X,+4,+6,+21
: A near-haploid karyotype with two copies of chromosomes 4, 6, and 21, and a single copy of all other chromosomes.

71,XXX,+8,+10
: A near-triploid karyotype with four copies of chromosomes 8 and 10, and three copies of all other chromosomes.

89,XXYY,–1,–3,–5,+8,–21
: A near-tetraploid karyotype with three copies of chromosomes 1, 3, 5, and 21, five copies of chromosome 8, and four copies of all other autosomes.

mos 47,XY,+21[12]/46,XY[18]
: A mosaic karyotype showing two cell lines, one cell line, represented by 12 cells, with trisomy 21 and one normal male cell line, represented by 18 cells. The normal diploid karyotype is written last.

The investigator should select as the reference for the description of the karyotype what is convenient and at the same time biologically meaningful. In such instances, the ploidy level (n, 2n, 3n, etc.) should be given in **angle brackets** < > after the chromosome number.

76~102<4n>,XXXX,...
: The chromosome numbers vary between hypertriploidy and hypertetraploidy. The symbol <4n> indicates that all abnormalities are expressed in relation to the tetraploid level.

58<2n>,XY,+X,+4,+6,+8,+10,+11,+14,+14,+17,+18,+21,+21[10]
: Near-triploid clone described in relation to the diploid chromosome number with gain of chromosomes listed.

8.2 Sex Chromosome Abnormalities

Constitutional sex chromosome abnormalities are described as follows:

45,X
    A karyotype with one X chromosome (Turner syndrome).

47,XXY
    A karyotype with two X chromosomes and one Y chromosome (Klinefelter syndrome).

47,XXX
    A karyotype with three X chromosomes.

47,XYY
    A karyotype with one X chromosome and two Y chromosomes.

48,XXXY
    A karyotype with three X chromosomes and one Y chromosome.

mos 47,XXY[10]/46,XY[20]
    A mosaic karyotype with one cell line showing two X chromosomes and one Y, found in 10 cells, and a second cell line with a normal diploid male pattern of one X chromosome and one Y chromosome, found in 20 cells.

mos 45,X[25]/47,XXX[12]/46,XX[13]
    A mosaic karyotype with two abnormal cell lines, one with monosomy X, found in 25 cells, and one with trisomy X, found in 12 cells. A normal female karyotype was found in 13 cells.

mos 47,XXX[25]/45,X[12]/46,XX[13]
    A mosaic karyotype with two abnormal cell lines, one with trisomy X found in 25 cells, and one with monosomy X found in 12 cells. A normal female karyotype was found in 13 cells.

mos 45,X[13]/46,XY[17]
    A mosaic karyotype with one cell line showing one X chromosome, found in 13 cells, and a second cell line with a normal diploid male pattern of one X chromosome and one Y chromosome, found in 17 cells.

45,X[15]/47,XXY[15]
    A mosaic karyotype with one cell line showing one X chromosome, found in 15 cells, and a second cell line showing two X chromosomes and one Y, found in 15 cells. When the number of abnormal cells is equivalent, losses are listed before gains. The use of the abbreviation mos is optional.

The constitutional sex chromosome complement is given without the use of plus or minus signs.

Acquired sex chromosome abnormalities are expressed with plus and minus signs as follows:

47,XX,+X
    A tumor karyotype in a female with an additional X chromosome.

45,X,–X
: A tumor karyotype in a female with loss of one X chromosome.

45,X,–Y
: A tumor karyotype in a male with loss of the Y chromosome.

45,Y,–X
: A tumor karyotype in a male with loss of the X chromosome.

48,XY,+X,+Y
: A tumor karyotype in a male with one additional X and one additional Y chromosome.

Acquired chromosome abnormalities in individuals with a constitutional sex chromosome anomaly can be distinguished with the use of the letter **c** after the constitutional abnormality designation, as illustrated in more detail in Section 11.3.

48,XXYc,+X
: Tumor cells with an acquired additional X chromosome in a patient with Klinefelter syndrome.

46,Xc,+X
: Tumor cells with an acquired additional X chromosome in a patient with Turner syndrome.

46,XXYc,–X
: Tumor cells with an acquired loss of one X chromosome in a patient with Klinefelter syndrome.

44,Xc,–X
: Tumor cells with an acquired loss of the X chromosome in a patient with Turner syndrome.

46,Xc,+21
: Tumor cells with an acquired extra chromosome 21 in a patient with Turner syndrome.

47,XXX?c
: Tumor cells with an uncertain karyotype with an extra X chromosome. The question mark indicates that it is unclear if the extra X is constitutional or acquired.

48,XXY,+mar c
: For constitutional markers, there is a space between **mar** and **c**.

8.3 Autosomal Abnormalities

Constitutional and acquired gains or losses of chromosomes are indicated with plus or minus signs.

47,XX,+21
: A karyotype with trisomy 21.

48,XX,+13,+21
: A karyotype with trisomy 13 and trisomy 21.

45,XX,–22
: A karyotype with monosomy 22.

46,XX,+8,–21
: A karyotype with trisomy 8 and monosomy 21.

Acquired autosomal abnormalities in individuals with a constitutional anomaly are, as exemplified above for sex chromosome abnormalities, distinguished by the letter **c** (see Section 11.3) after the constitutional abnormality designation.

48,XY,+21c,+21
: An acquired extra chromosome 21 in a patient with Down syndrome.

46,XY,+21c,–21
: Acquired loss of one chromosome 21 in a patient with Down syndrome.

9 Structural Chromosome Rearrangements

9.1 General Principles

Structural aberrations, whether constitutional or acquired, should be expressed in relation to the appropriate ploidy level (see Section 11.2), i.e., in near-haploid cells in relation to one chromosome of each type, in near-diploid cells in relation to two chromosomes of each type, in near-triploid cells in relation to three chromosomes of each type, in near-tetraploid cells in relation to four chromosomes of each type, and so on.

69,XXX,del(7)(p11.2)
: Two normal chromosomes 7 and one with a deletion of the short arm.

69,XXY,del(7)(q22),inv(7)(p13q22),t(7;14)(p15;q11.1)
: No normal chromosome 7: one has a long-arm deletion, one has an inversion, and one is involved in a balanced translocation with chromosome 14.

70,XXX,+del(7)(p11.2)
: Three normal chromosomes 7 and an additional, structurally abnormal chromosome 7 with a deletion of the short arm.

92,XXYY,del(7)(p11.2),t(7;14)(p15;q11.1)
: Two normal and two abnormal chromosomes 7: one has a deletion of the short arm, and one is involved in a balanced translocation with chromosome 14.

92,XXYY,del(7)(p11.2),del(7)(q22),del(7)(q34)
: One normal chromosome 7 and three with different deletions.

When normal chromosomes are replaced by structurally altered chromosomes, the normal ones should not be recorded as missing.

46,XX,inv(3)(q21q26.2)
: An inversion of one chromosome 3. There is no need to indicate that one chromosome 3 is missing, i.e., the karyotype should not be written 46,XX,−3,+inv(3).

45,XX,dic(13;15)(q22;q24)
: It is apparent from the symbol dic (see Section 9.2.4) and from the specification of the chromosomes involved that the dicentric chromosome replaces two normal chromosomes. Thus, there is no need to indicate the missing normal chromosomes.

46,Y,t(X;8)(p22.3;q24.1)
: Male karyotype showing a balanced translocation between the X chromosome and chromosome 8. Note that the normal sex chromosome, in this case a Y, is shown first.

46,XY,der(1)t(1;3)(p22;q13.1)
: The der(1) replaces a normal chromosome 1 and there is no need to indicate the missing normal chromosome. The description implies that the karyotype contains one normal chromosome 1 and two normal chromosomes 3.

46,XX,der(1)ins(1;?)(p22;?)
: Material of unknown origin has been inserted at band p22 in one chromosome 1. The homologous chromosome 1 is normal.

45,XY,−10,der(10)t(10;17)(q22;p12)
: The der(10) replaces a normal chromosome 10; the homologous chromosome 10 is lost. In this situation the missing chromosome 10 must be indicated.

9.2 Specification of Structural Rearrangements

Examples of structural rearrangements, whether constitutional or acquired, are presented below. Each abnormality is described first with the short form and, when appropriate, also with the detailed form.

9.2.1 Additional Material of Unknown Origin

The symbol **add** (Latin, *additio*) should be used to indicate additional material of unknown origin attached to a chromosome region or band. Such abnormalities have often been described using the symbols t and ?, e.g., t(1;?)(p36;?), but it is only rarely known that the rearranged chromosome has actually resulted from a translocation. The symbol add does not imply any particular mechanism and is therefore recommended.

Additional material attached to a terminal band will always lead to an increase in length of a chromosome arm. Unknown material that replaces a chromosome segment may, depending on the size of the extra material, result in either increase or decrease in the length of the chromosome arm. Designations such as "1p+" or "1p−" may be used in text to describe such abnormal chromosomes, but should not be used in the karyotype.

46,XX,add(19)(p13.3)
46,XX,add(19)(?::p13.3→qter)
: Additional material attached to band 19p13.3, but neither the origin of the extra segment nor the type of rearrangement is known.

46,XY,add(12)(q13)
46,XY,add(12)(pter→q13::?)
: Additional material of unknown origin replaces the segment 12q13qter.

When additional material of unknown origin is attached to both arms of a chromosome and/or replaces more than one segment in a chromosome, the symbol **der** (see Section 9.2.3) should be used.

46,XX,der(5)add(5)(p15.3)add(5)(q23)
46,XX,der(5)(?::p15.3→q23::?)
    Additional material of unknown origin is attached at band 5p15.3 in the short arm and additional material replaces the segment 5q23qter in the long arm.

Unknown material *inserted* in a chromosome arm should be described by the use of the symbols **ins** and **?**.

46,XX,der(5)ins(5;?)(q13;?)
46,XX,der(5)ins(5;?)(pter→q13::?::q13→qter)
    Material of unknown origin has been inserted into the long arm of chromosome 5 at band 5q13. Use of the symbol add in this situation, i.e., add(5)(q13), would have denoted that unknown material had replaced the segment 5q13qter.

9.2.2 Deletions

The symbol **del** is used to denote both *terminal* and *interstitial deletions*. A deletion is a loss of a chromosome segment. No arrows are used in the short form to indicate the extent of the deleted segment. This is apparent from the description of the breakpoints. Note that designations such as "5q–" or "del(5q)", which may be useful abbreviations in text, should not be used in karyotypes.

46,XX,del(5)(q13)
46,XX,del(5)(pter→q13:)
    Terminal deletion with a break (:) in band 5q13. The remaining chromosome consists of the entire short arm of chromosome 5 and the part of the long arm lying between the centromere and band 5q13.

46,XX,del(4)(p15.2)
46,XX,del(4)(:p15.2→qter)
    Terminal deletion with a break (:) in band 4p15.2. The remaining chromosome consists of the part of the short arm of chromosome 4 between band 4p15.2 and the centromere and the entire long arm.

46,XX,del(5)(q13q33)
46,XX,del(5)(pter→q13::q33→qter)
    Interstitial deletion with breakage and reunion (::) of bands 5q13 and 5q33. The segment lying between these bands has been deleted.

46,XX,del(5)(q13q13)
46,XX,del(5)(pter→q13::q13→qter)
    Interstitial deletion of a small segment within band 5q13, i.e., both breakpoints are in band 5q13.

46,XY,del(5)(q?)
    Deletion of the long arm of chromosome 5, but it is unclear whether it is a terminal or an interstitial deletion, and also the breakpoints are unknown.

46,Y,del(X)(p21p11.4)
    Interstitial deletion of the segment between bands Xp21 and Xp11.4.

46,XY,del(20)(q11.2–13.1q13.3)
    Interstitial deletion identified at 300-band resolution with breakpoints at 20q11.2–13.1 and 20q13.3.

Multiple deletions of the same chromosome should be expressed using the symbol **der** (see Section 9.2.3).

9.2.3 Derivative Chromosomes

A *derivative chromosome* (**der**) is a structurally rearranged chromosome generated either by a rearrangement involving two or more chromosomes or by multiple aberrations within a single chromosome. The term always refers to the chromosome that has an intact centromere.

A *recombinant chromosome* (**rec**) is a structurally rearranged chromosome with a new segmental composition resulting from *meiotic* crossing-over; consequently, this term should not be used in the description of acquired chromosome abnormalities, nor those resulting from malsegregation. If parental karyotypes are unknown or a parental inversion has not been identified, the abnormal chromosome should be designated as a der, not a rec.

46,XX,der(6)(pter→q25.2::p22.2→pter)

When parental karyotypes are known and a parental inversion or an intra- or interchromosomal insertion is identified, rec should be used.

46,XX,rec(6)dup(6p)inv(6)(p22.2q25.2)dmat
46,XX,rec(6)(pter→q25.2::p22.2→pter)dmat
    Recombinant chromosome 6 containing a duplication of segment 6p22.2 to 6pter and a deletion of 6q25.2 to 6qter due to a meiotic crossing-over in the mother. The mother is carrier of an inversion of the segment 6p22.2 to 6q25.2.

46,XX,rec(21)del(21)ins(21)(p13q22.2q22.3)dpat
    Recombinant chromosome 21 containing a deletion of segment 21q22.2q22.3 due to a meiotic crossing-over in the father. The father is carrier of an intrachromosomal insertion of the bands 21q22.2 to 21q22.3 into p13.

46,XY,rec(1)dup(5q)ins(1;5)(q32;q11.2q22)dinh,rec(5)del(1q)ins(1;5)dinh
    Recombinant chromosomes 1 and 5 resulting in a duplication of segment 5q22 to 5qter and deletion of the segment from 1q32 to 1qter due to a meiotic crossing-over in a parent who carries an interchromosomal insertion of the segment from 5q11.2 to 5q22 into band 1q32.

A **derivative chromosome generated by more than one rearrangement within a chromosome** is specified in parentheses, followed by the type of abnormality. The detailed form is helpful in these cases.

46,XY,der(9)del(9)(p12)del(9)(q31)
46,XY,der(9)(:p12→q31:)
    A derivative chromosome 9 resulting from terminal deletions in both the short and long arms with breakpoints in bands 9p12 and 9q31.

46,XY,der(9)inv(9)(p23p13)del(9)(q22q33)
46,XY,der(9)(pter→p23::p13→p23::p13→q22::q33→qter)
    A derivative chromosome 9 resulting from an inversion in the short arm with breakpoints in 9p23 and 9p13, and an interstitial deletion of the long arm with breakpoints in 9q22 and 9q33.

46,XX,der(7)add(7)(p22)add(7)(q22)
46,XX,der(7)(?::p22→q22::?)
    A derivative chromosome 7 with additional material of unknown origin attached at band 7p22. Similarly, additional material of unknown origin is attached to 7q22, replacing the segment 7q22qter.

46,XX,der(5)add(5)(p15.1)del(5)(q13)
46,XX,der(5)(?::p15.1→q13:)
A derivative chromosome 5 with additional material of unknown origin attached at 5p15.1 and a terminal deletion of the long arm distal to band 5q13.

A derivative chromosome resulting from one rearrangement involving two or more chromosomes is specified in parentheses, followed by the type of abnormality.

46,Y,der(X)t(X;8)(p22.3;q24.1)
A male showing a derivative X chromosome derived from a translocation between Xp22.3 and 8q24.1.

46,XX,der(1)t(1;3)(p22;q13.1)
46,XX,der(1)(3qter→3q13.1::1p22→1qter)
The derivative chromosome 1 has resulted from a translocation of the chromosome 3 segment distal to 3q13.1 to the short arm of chromosome 1 at band 1p22. The der(1) replaces a normal chromosome 1 and there is no need to indicate the missing chromosome (see Section 9.1). There are two normal chromosomes 3. The karyotype is unbalanced with loss of the segment 1p22pter and gain of 3q13.1qter.

45,XY,der(1)t(1;3)(p22;q13.1),–3
The derivative chromosome 1 (same as above) replaces a normal chromosome 1, but there is only one normal chromosome 3. One can presume that it is the der(3) resulting from the t(1;3) that has been lost, but the karyotype cannot make explicit such assumptions.

47,XX,+7,der(7)t(1;7)(q12;p22)×2
In addition to one normal copy of chromosome 7, two copies of the derivative chromosome 7 from a translocation between chromosomes 1 and 7 with breakpoints at 1q12 and 7p22 are present. There are two normal copies of chromosome 1.

47,XY,+der(4)t(4;11)(q21;q23),t(4;11)(q21;q23)
In addition to the balanced translocation between the long arms of chromosomes 4 and 11, there is an additional copy of the derivative chromosome 4 from this translocation. The derivative chromosome is listed before the translocation.

The term *Philadelphia chromosome* is for historical reasons retained to describe the derivative chromosome 22 generated by the translocation t(9;22)(q34;q11.2). The abbreviation **Ph** (formerly Ph[1]) may be used in text, but not in the description of the karyotype, where der(22)t(9;22)(q34;q11.2) is recommended. Similarly, the derivative chromosome 9 resulting from the t(9;22) is designated der(9)t(9;22)(q34;q11.2).

A derivative chromosome generated by more than one rearrangement involving two or more chromosomes is specified in parentheses, followed by all aberrations involved in the generation of the derivative chromosome. The aberrations should be listed according to the breakpoints of the derivative chromosome from pter to qter and should not be separated by commas.

46,XX,der(1)t(1;3)(p32;q21)t(1;11)(q25;q13)
46,XX,der(1)(3qter→3q21::1p32→1q25::11q13→11qter)
A derivative chromosome 1 generated by two translocations, one involving the short arm with a breakpoint in 1p32 and the other involving the long arm with a breakpoint in 1q25.

46,XY,der(1)t(1;3)(p32;q21)t(3;7)(q28;q11.2)
46,XY,der(1)(7qter→7q11.2::3q28→3q21::1p32→1qter)
   A derivative chromosome 1 resulting from a translocation of the chromosome 3 segment distal to 3q21 onto 1p32, and a translocation of the segment 7q11.2qter to band 3q28 of the chromosome 3 segment attached to chromosome 1.

46,XY,der(1)t(1;3)(p32;q21)dup(1)(q25q42)
46,XY,der(1)(3qter→3q21::1p32→1q42::1q25→1qter)
   A derivative chromosome 1 resulting from a t(1;3) with a breakpoint in 1p32 and a duplication of the long arm segment 1q25q42.

46,XY,der(9)del(9)(p12)t(9;13)(q34;q11)
46,XY,der(9)(:9p12→9q34::13q11→13qter)
   A derivative chromosome 9 generated by a terminal deletion of the short arm with a breakpoint in 9p12, and by a t(9;13) involving the long arm with a breakpoint in 9q34.

46,XX,der(1)t(1;11)(p32;q13)t(1;3)(q25;q21)
46,XX,der(1)(11qter→11q13::1p32→1q25::3q21→3qter)
   A derivative chromosome 1 generated by two translocations, one involving a breakpoint in 1p32 and 11q13 and the other involving a breakpoint in 1q25 and 3q21. The detailed form describes the derivative 1 from 11qter to 3qter as the aberrations are listed according to the orientation of chromosome 1, from the p arm to the q arm.

47,XY,+der(8)r(1;8;17)(p36.3p35;p12q13;q25q25)
47,XY,+der(8)r(1;8;17)(::1p36.3→1p35::8p12→8q13::17q25→17q25::)
   A supernumerary ring chromosome involving three chromosome segments determined to be a derivative 8 because it retains the 8 centromere. For additional examples of ring chromosomes, see 9.2.15.

46,XX,der(1)del(1)(p34p22)ins(1;17)(p34;q25q11)
46,XX,der(1)(1pter→1p34::17q25→17q11::1p22→1qter)
   A derivative chromosome 1 resulting from an interstitial deletion of the short arm with breakpoints in 1p34 and 1p22, and a replacement of this segment by an insertion of a segment from the long arm of chromosome 17. In such situations, when there are two breakpoints in the recipient chromosome, the breakpoint closest to pter is listed as the point of insertion. The orientation of the inserted segment relative to pter and qter of the derivative chromosome has been reversed in its new position.

46,XY,der(7)t(2;7)(q21;q22)ins(7;?)(q22;?)
46,XY,der(7)(7pter→7q22::?::2q21→2qter)
   A derivative chromosome 7 in which material of unknown origin has replaced the segment 7q22qter, and the segment 2q21qter from the long arm of chromosome 2 is attached to the unknown chromosome material. By convention, the breakpoint in the derivative chromosome is specified as the point of insertion of the unknown material.

47,XX,t(9;22;6)(q34;q11.2;p21),+der(22)t(9;22;6)
   A three-way translocation between chromosomes 9, 22, and 6 is present along with an additional derivative chromosome 22 from this three-way translocation.

46,XX,der(8)t(8;17)(p23;q21)inv(8)(p22q13)t(8;22)(q22;q12)
46,XX,der(8)(22qter→22q12::8q22→8q13::8p22→8q13::8p22→8p23::17q21→17qter)
   A derivative chromosome 8 resulting from two translocations, one affecting the short arm, one the long arm, with breakpoints at 8p23 and 8q22, respectively, and a pericentric inversion with breakpoints at 8p22 and 8q13.

An *isoderivative chromosome*, abbreviated **ider**, designates an isochromosome formation for one of the arms of a derivative chromosome. The breakpoints are assigned to the centromeric bands p10 and q10 according to the morphology of the isoderivative chromosome (see Section 9.2.11).

46,XX,ider(22)(q10)t(9;22)(q34;q11.2)
46,XX,ider(22)(9qter→9q34::22q11.2→22q10::22q10→22q11.2::9q34→9qter)
   An isochromosome for the long arm of a derivative chromosome 22 generated by a t(9;22), i.e., an isochromosome for the long arm of a Ph chromosome.

46,XY,ider(9)(p10)ins(9;12)(p13;q22q13)
46,XY,ider(9)(9pter→9p13::12q22→12q13::9p13→9p10::9p10→9p13::12q13→12q22::9p13→9pter)
   An isochromosome for the short arm of a derivative chromosome 9 resulting from an insertion of the segment 12q13q22 at band 9p13 with band 12q22 closer to 9pter than band 12q13.

When a derivative chromosome is dicentric and contains one or more additional abnormalities, the two centromere-containing chromosomes are given within parentheses, separated by a semicolon, followed by the specification of the aberrations.

45,XX,der(5;7)t(5;7)(q22;p13)t(3;7)(q21;q21)
45,XX,der(5;7)(5pter→5q22::7p13→7q21::3q21→3qter)
   A dicentric derivative chromosome. Breakage and reunion have occurred at band 5q22 in the long arm of chromosome 5 and at band 7p13 in the short arm of chromosome 7. In addition, the segment 3q21qter has been translocated onto the long arm of chromosome 7 at band 7q21.

45,XY,der(5;7)t(3;5)(q21;q22)t(3;7)(q29;p13)
45,XY,der(5;7)(5pter→5q22::3q21→3q29::7p13→7qter)
   A dicentric derivative chromosome composed of chromosomes 5 and 7. The same acentric chromosome 3 segment as in the previous example is inserted between the long arm of chromosome 5 and the short arm of chromosome 7.

45,XY,der(5;7)t(3;5)(q21;q22)t(3;7)(q29;p13)del(7)(q32)
45,XY,der(5;7)(5pter→5q22::3q21→3q29::7p13→7q32:)
   The same dicentric derivative chromosome as in the previous example but with an additional terminal deletion of the long arm of chromosome 7 at band 7q32.

45,XX,der(8;8)(q10;q10)del(8)(q22)t(8;9)(q24.1;q12)
45,XX,der(8;8)(:8q22→8q10::8q10→8q24.1::9q12→9qter)
   A derivative chromosome composed of the long arms of chromosome 8 with material from chromosome 9 translocated to one arm at band 8q24.1 and a deletion at band 8q22 in the other arm.

46,XY,der(9)t(9;22)(q34;q11),+22,der(22;22)(22pter→22q11::9q34→9qter::9qter→9q34::22q11→22pter)
   An isochromosome for the derivative chromosome 22 generated by a t(9;22), i.e., an isochromosome for the long arm of a Ph chromosome. There are also two normal copies of chromosome 22 and the derivative chromosome 9 generated by a t(9;22).

When the centromere of the derivative chromosome is not known, but more distal parts of the chromosome can be recognized, the abnormal chromosome may be designated **der(?)**.

47,XY,+der(?)t(?;9)(?;q22)
47,XY,+der(?)(?→?cen→?::9q22→9qter)
    The distal segment of the long arm of chromosome 9 from band 9q22 has been translocated to a centromere-containing derivative chromosome of unknown origin.

47,XX,+der(?)t(?;9)(?;p13)ins(?;7)(?;q11.2q32)
47,XX,+der(?)(9pter→9p13::?→cen→?::7q11.2→7q32::?)
    A derivative chromosome of unknown origin onto which is translocated in its short arm the segment of chromosome 9 distal to band 9p13, and which also contains an insertion in the long arm of the chromosome 7 segment between bands 7q11.2 and 7q32.

47,XX,+der(?)t(?;9)(?;p13)hsr(?)
47,XX,+der(?)(9pter→9p13::?→cen→?::hsr→?)
    A derivative chromosome of unknown origin with the same translocation in its short arm as in the previous example, and a homogeneously staining region in the long arm.

Derivative chromosomes whose centromeres are unknown should be placed after all identified abnormalities but before unidentified ring chromosomes, marker chromosomes, and double minute chromosomes (see Chapter 6).

53,XX,…,+der(?)t(?;9)(?;q22),+r,+mar,dmin

There is usually no need to indicate which homologue is involved in a derivative chromosome because this will be apparent from the karyotype description. If both homologues are involved, this will result in two derivative homologous chromosomes.

46,XX,der(9)del(9)(p12)t(9;22)(q34;q11.2),der(9)t(9;12)(p13;q22)inv(9)(q13q22)
    One der(9) is the result of a deletion of the short arm and a translocation involving the long arm; the other der(9) is the result of a translocation affecting the short arm and a paracentric inversion in the long arm of the homologous chromosome 9. There are two normal chromosomes 12, two normal chromosomes 22, but no normal chromosome 9.

When homologous chromosomes cannot be distinguished within this nomenclature system, one of the numerals may be underlined (<u>single underlining</u>). There is in particular one situation where this may be helpful: When the two homologous chromosomes are involved in identical aberrations resulting in two identical derivative chromosomes.

46,XX,der(1)t(1;3)(p34.3;q21),der(<u>1</u>)t(<u>1</u>;3)(p34.3;q21)
    The two homologous chromosomes 1, as identified by C-band polymorphism, are involved in apparently identical translocations.

46,XX,der(1)t(1;3)(p34.3;q21)[20]/46,XX,der(<u>1</u>)t(<u>1</u>;3)(p34.3;q21)[10]
    The two homologous chromosomes 1 are involved in apparently identical translocations in different cells. The two abnormalities represent two different clones; the homologous chromosomes 1 in each clone are normal.

46,XX,der(<u>1</u>)t(1;<u>1</u>)(p31;q32)
46,XX,der(1)t(1;<u>1</u>)(p31;q32)
    The two derivative chromosomes can be distinguished by underlining. In the first example, the derivative observed is the homologue with a break at q32. In the second example, the derivative is the homologue with a breakpoint in p31.

Complex rearrangements may give rise to several derivative chromosomes. The breakpoints in the derivative chromosomes generated by the *same rearrangement* need not be repeated in the description of each individual derivative chromosome.

47,XX,t(9;22)(q34;q11.2),+der(22)t(9;22)
    Karyotype with t(9;22) and an additional Ph chromosome. The breakpoints in the extra der(22) need not be repeated.

46,XX,der(1)t(1;3)(p32;q21)inv(1)(p22q21)t(1;11)(q25;q13),der(3)t(1;3),der(11)t(1;11)
    A balanced complex rearrangement with three derivative chromosomes. The breakpoints of the t(1;3) and the t(1;11), which both contribute to the der(1), are not repeated in the description of der(3) and der(11).

Complex karyotypes involving rearrangements between two or more derivative chromosomes, or where derivative chromosomes are involved in new rearrangements, cannot be described by the short form. The detailed form will be adequate in all such situations. It is acceptable to combine the short form (4.3.1) and the detailed form (4.3.2) for designating complex karyotypes. Whenever doubts remain, the rearrangement should, to avoid ambiguity, be illustrated and described in words.

9.2.4   Dicentric Chromosomes

The symbol **dic** is used to describe *dicentric chromosomes*. It is apparent from the symbol and from the specification of the chromosome(s) involved that the dicentric chromosome replaces two normal chromosomes. A dicentric chromosome is counted as one chromosome, with the normal chromosome count then becoming 45. There is no need to indicate the missing normal chromosome(s) (cf., whole-arm and Robertsonian translocations, Sections 9.2.17.2 and 9.2.17.3). Two breakpoints are specified and the centromeres are presumed to have arisen from the two separate chromosomes.

This is in contrast to the mechanism from which an isodicentric arises where there is a single breakpoint on sister chromatids and a subsequent reunion; the normal chromosome count is therefore 46. *Isodicentric chromosomes* are designated **idic**.

The term **der** may be used instead of **dic**, but the combination of **der dic** should never be used.

45,XX,dic(13;13)(q14;q32)
45,XX,dic(13;13)(13pter→13q14::13q32→13pter)
    Breakage and reunion have occurred at bands 13q14 and 13q32 on the two homologous chromosomes 13 to form a dicentric chromosome. There is no normal chromosome 13. If it can be shown that the dicentric chromosome has originated through breakage and reunion of sister chromatids, it may be designated, e.g., dic(13)(q14q32).

45,XX,dic(13;15)(q22;q24)
45,XX,dic(13;15)(13pter→13q22::15q24→15pter)
    A dicentric chromosome with breaks and reunion at bands 13q22 and 15q24. The missing chromosomes 13 and 15 are not indicated since they are replaced by the dicentric chromosome. The karyotype contains one normal chromosome 13, one normal chromosome 15, and the dic(13;15). The resulting net imbalance of this abnormality is loss of the segments distal to 13q22 and 15q24.

46,XX,+13,dic(13;15)(q22;q24)
: A dicentric chromosome with breaks and reunion at bands 13q22 and 15q24 (same as above) has replaced one chromosome 13 and one chromosome 15. There are, however, two normal chromosomes 13, i.e., an additional chromosome 13 in relation to the expected loss due to the dic(13;15). Consequently, the gain is indicated as +13. The karyotype contains two normal chromosomes 13, one normal chromosome 15, and the dic(13;15). The resulting net imbalance is partial trisomy for the segment 13pterq22 and loss of the segment 15q24qter.

45,XY,dic(14;21)(p11.2;p11.2)
45,XY,dic(14;21)(14qter→14p11.2::21p11.2→21qter)
: A dicentric chromosome with breaks and reunion at bands 14p11.2 and 21p11.2. The missing chromosomes 14 and 21 are not indicated since they are replaced by the dicentric chromosome. The karyotype contains one normal chromosome 14, one normal chromosome 21, and the dic(14;21). The resulting net imbalance of this abnormality is loss of the segments distal to 14p11.2 and 21p11.2. For description of Robertsonian translocations, see Section 9.2.17.3.

47,XY,+dic(17;?)(q22;?)
47,XY,+dic(17;?)(17pter→17q22::?)
: An additional dicentric chromosome composed of one chromosome 17 with a break at band 17q22 and an unknown chromosome with an intact centromere.

46,X,idic(Y)(q12)
46,X,idic(Y)(pter→q12::q12→pter)
: Breakage and reunion have occurred at band Yq12 on sister chromatids to form an isodicentric Y chromosome. The resulting net imbalance is loss of the segment Yq12qter and gain of Ypterq12.

46,XX,idic(21)(q22.3)
46,XX,idic(21)(pter→q22.3::q22.3→pter)
: An isodicentric with breakage and reunion at the terminal ends of two chromosomes 21. There are two copies of the long arm of chromosome 21, joined at q22.3, and one normal chromosome 21, indicated by the 46 count. Even though there are effectively three copies of the chromosome 21 long arm, the normal chromosome 21 is not designated with a (+) sign.

47,XX,+idic(13)(q22)
47,XX,+idic(13)(pter→q22::q22→pter)
: An additional isodicentric chromosome 13. There are two chromosomes 13 and the idic(13). Another example is shown in Section 9.2.11.

47,XY,+idic(15)(q12)
47,XY,+dic(15;15)(q12;q12)
47,XY,+dic(15;15)(pter→q12::q12→pter)
: An additional apparent isodicentric chromosome 15. There are two chromosomes 15 and the idic(15)(q12). This rearrangement has historically been referred to as inv dup(15)(q12). However, because most result from recombination between homologues, dic(15;15)(q12;q12), (or psu dic, see below), would be a more appropriate designation.

Complex dicentric chromosomes must be described as derivative chromosomes, see Section 9.2.3.

A *pseudodicentric chromosome* is a dicentric structure in which only one centromere is active. Such chromosomes are abbreviated **psu dic** (similarly, *pseudotricentric*, **psu trc**, etc.), and the segment with the presumptively active centromere, based on the morphology in the majority of cells, is always written first. If the active centromere cannot be determined, the smallest chromosome number is written first.

45,XX,psu dic(15;13)(q12;q12)
45,XX,psu dic(15;13)(15pter→15q12::13q12→13pter)
    A pseudodicentric chromosome has replaced one chromosome 13 and one chromosome 15. The karyotype contains one normal chromosome 13, one normal chromosome 15, and the psu dic(15;13). The centromere of the chromosome stated first, i.e., chromosome 15, is the active one.

46,XX,psu idic(20)(q11.2)
46,XX,psu idic(20)(pter→q11.2::q11.2→pter)
    A pseudodicentric chromosome has replaced one chromosome 20, resulting in three copies of 20pterq11.2. The psu idic(20) has one active centromere.

9.2.5 Duplications

The symbol **dup** indicates a *duplication*. Duplications are a gain of a chromosome segment observed at the original chromosome location. When a gain of a chromosome segment is found elsewhere in the genome, **der** or **ins** should be used depending on the rearrangement. The orientation of the duplicated segment is indicated by the order of the bands from pter to qter. Note that no arrow is used in the short form to indicate the orientation.

46,XX,dup(1)(p34p31)
46,XX,dup(1)(pter→p31::p34→qter)
    Duplication of the segment between bands 1p34 and 1p31 with no change in orientation.

46,XX,dup(1)(p31p34)
46,XX,dup(1)(pter→p31::p31→p34::p31→qter) or dup(1)(pter→p34::p31→p34::p31→pter)
    Duplication of the segment between bands 1p34 and 1p31, in reversed orientation relative to pter and qter. Note that only the detailed form will clarify the location of the duplicated segment.

46,XX,dup(1)(q22q25)
46,XX,dup(1)(pter→q25::q22→qter)
    Duplication of the segment between bands 1q22 and 1q25.

46,XY,dup(1)(q25q22)
46,XY,dup(1)(pter→q25::q25→q22::q25→qter) or dup(1)(pter→q22::q25→q22::q22→qter)
    Duplication of the segment between bands 1q22 and 1q25, in reversed orientation relative to pter and qter. Note that only the detailed form will clarify the location of the duplicated segment.

9.2.6 Fission

The symbol **fis** is used to denote *centric fission*.

47,XY,−10,+fis(10)(p10),+fis(10)(q10)
47,XY,−10,+fis(10)(pter→p10:),+fis(10)(qter→q10:)
    Break in the centromere resulting in two derivative chromosomes composed of the short and long arms, respectively. The breakpoints (:) are assigned to p10 and q10 according to the morphology of the derivative chromosomes.

9.2.7 Fragile Sites

Fragile sites, abbreviated **fra**, may occur as normal variants (see Section 7.2) or be associated with specific diseases and/or phenotypic abnormalities. In both situations the same nomenclature is used.

46,X,fra(X)(q27.3)
    A fragile site in sub-band Xq27.3 on one X chromosome in a female.

46,Y,fra(X)(q27.3)
    A fragile site in sub-band Xq27.3 on the X chromosome in a male.

45,fra(X)(q27.3)
    A fragile site in sub-band Xq27.3 on the X chromosome in a Turner syndrome patient.

47,XY,fra(X)(q27.3)
    A fragile site in sub-band Xq27.3 on one X chromosome in a Klinefelter syndrome patient.

9.2.8 Homogeneously Staining Regions

The symbol **hsr** is used to describe the presence, but not the size, of *a homogeneously staining region* in a chromosome arm, segment, or band.

46,XX,hsr(1)(p22)
46,XX,hsr(1)(pter→p22::hsr::p22→qter)
    A homogeneously staining region in band 1p22.

46,XY,hsr(21)(q22)
46,XY,hsr(21)(pter→q22::hsr::q22→qter)
    A homogeneously staining region in band 21q22.

When a chromosome contains multiple hsr or one hsr and another structural change, it is by definition a derivative chromosome and should be designated accordingly (see Section 9.2.3).

46,XX,der(1)hsr(1)(p22)hsr(1)(q31)
46,XX,der(1)(pter→p22::hsr::p22→q31::hsr::q31→qter)
    Two homogeneously staining regions in chromosome 1: one in band 1p22 in the short arm and the other in band 1q31 in the long arm.

46,XY,der(1)del(1)(p33p21)hsr(1)(p21)
46,XY,der(1)(pter→p33::hsr::p21→qter)
    The segment between bands 1p21 and 1p33 is replaced by a homogeneously staining region that may be smaller or larger than the deleted segment. The hsr is by convention assigned to the proximal deletion breakpoint band.

When a homogeneously staining region is located at the interface between segments of different chromosomes involved in a rearrangement, the hsr is assigned to the breakpoints in both chromosomes according to the standard nomenclature for structural chromosome aberrations, i.e., the two chromosomes involved are presented in the first parentheses and the breakpoints in the second.

46,XX,der(1)ins(1;7)(q21;p21p11.2)hsr(1;7)(q21;p11.2)
46,XX,der(1)(1pter→1q21::hsr::7p11.2→7p21::1q21→1qter)
    Insertion of the segment 7p21p11.2 into the long arm of chromosome 1 with breakage and reunion at band 1q21. The derivative chromosome also contains an hsr at the interface between the recipient and donor chromosomes. The hsr is located proximal to the segment inserted from chromosome 7.

46,XX,der(1)ins(1;7)(q21;p21p11.2)hsr(1;7)(q21;p21)
46,XX,der(1)(1pter→1q21::7p11.2→7p21::hsr::1q21→1qter)
    Insertion of the segment 7p21p11.2 into the long arm of chromosome 1 with breakage and reunion at band 1q21. The derivative chromosome also contains an hsr at the interface between the recipient and donor chromosomes. The hsr is located distal to the segment inserted from chromosome 7.

The distinction between homogeneously staining regions and abnormally banded regions, another term that has been used to describe regions containing amplified genes, is ambiguous. So is the distinction between abnormally banded regions and any region of unknown origin within a derivative or marker chromosome. Therefore, until better defined, no symbol to denote abnormally banded regions should be used in karyotype descriptions.

9.2.9 Insertions

The symbol **ins** is used for *insertions*. The orientation of the inserted segment is indicated by the order of the bands of the inserted segment with respect to the centromere. For clarity, the use of the detailed form is recommended.

Insertion within a chromosome

46,XX,ins(2)(p13q31q21)
46,XX,ins(2)(pter→p13::q31→q21::p13→q21::q31→qter)
    The long-arm segment between bands 2q21 and 2q31 has been inserted into the short arm at band 2p13. The original orientation of the inserted segment relative to pter and qter has been reversed in its new position, i.e., band 2q31 is now closer to pter than band 2q21.

46,XY,ins(2)(p13q21q31)
46,XY,ins(2)(pter→p13::q21→q31::p13→q21::q31→qter)
    The insertion is the same as in the previous example except that the orientation of the bands within the segment has been maintained relative to pter and qter, i.e., band 2q21 of the inserted segment remains closer to pter than 2q31.

Insertion between two chromosomes

46,XX,ins(5;2)(p14;q32q22)
46,XX,ins(5;2)(5pter→5p14::2q32→2q22::5p14→5qter;2pter→2q22::2q32→2qter)
    The long-arm segment between bands 2q22 and 2q32 has been inserted into the short arm of chromosome 5 at band 5p14. The original orientation of the inserted segment relative to pter and qter has been reversed in its new position, i.e., 2q32 is now closer to pter of the recipient chromosome than 2q22. Note that the recipient chromosome is specified first.

46,XY,ins(5;2)(p14;q22q32)
46,XY,ins(5;2)(5pter→5p14::2q22→2q32::5p14→5qter;2pter→2q22::2q32→2qter)
   Breakage and reunion have occurred at the same bands as in the previous example except that band 2q22 remains closer to pter of the recipient chromosome than band 2q32.

46,XX,ins(5;2)(q31;p13p23)
46,XX,ins(5;2)(5pter→5q31::2p13→2p23::5q31→5qter;2pter→2p23::2p13→2qter)
   An insertion of bands p23 to p13 from chromosome 2 into band 5q31 with reversal of orientation relative to pter of the recipient chromosome 5.

46,XX,ins(5;2)(q31;p23p13)
46,XX,ins(5;2)(5pter→5q31::2p23→2p13::5q31→5qter;2pter→2p23::2p13→2qter)
   An insertion of bands p23 to p13 from chromosome 2 into band 5q31 in a reverse orientation to the above example.

46,XY,der(5)ins(5;2)(q31;p23p13)dmat
   A derivative chromosome 5 resulting from malsegregation of a maternal insertion. There is one derivative chromosome 5 containing the insertion from chromosome 2, one normal chromosome 5, and two normal chromosomes 2.

46,X,der(X)ins(X;7)(p21;q22q21)
   A derivative X chromosome resulting from an insertion of the segment from 7q21 to 7q22 into band Xp21, with band 7q22 closer to Xpter than band 7q21. There are two normal chromosomes 7.

9.2.10 Inversions

The symbol **inv** is used. Whether it is a *paracentric* or *pericentric inversion* is apparent from the band designations. In all cases, the breakpoint closer to pter of the inverted chromosome is specified first.

46,XX,inv(2)(p23p13)
   Paracentric inversion in which breakage and reunion have occurred at bands 2p13 and 2p23.

46,XX,inv(3)(q21q26.2)
46,XX,inv(3)(pter→q21::q26.2→q21::q26.2→qter)
   Paracentric inversion in which breakage and reunion have occurred at bands 3q21 and 3q26.2.

46,XY,inv(3)(p13q21)
46,XY,inv(3)(pter→p13::q21→p13::q21→qter)
   Pericentric inversion in which breakage and reunion have occurred at bands 3p13 and 3q21. The breakpoint in the short arm is specified first.

9.2.11 Isochromosomes

The symbol **i** (not iso) is used for *isochromosomes* and **idic** for *isodicentric chromosomes*. The breakpoints in isochromosomes are assigned to the centromeric bands p10 and q10. The isochromosome designation is inferred from the banded chromosome morphology. Isochromosomes are usually formed through a centric mis-division and result in chromosome arms which are a mirror image of each other and genetically homozygous. See also Section 9.2.4.

Complex isochromosomes, including isoderivative chromosomes, are described as derivative chromosomes, see Section 9.2.3.

46,XX,i(17)(q10)
46,XX,i(17)(qter→q10::q10→qter)
An isochromosome for the entire long arm of one chromosome 17 and consequently the breakpoint is assigned to 17q10. There is one normal chromosome 17. The shorter designation i(17q) may be used in text but not in the karyotype to describe this isochromosome.

46,X,i(X)(q10)
46,X,i(X)(qter→q10::q10→qter)
One normal X chromosome and an isochromosome for the long arm of one X chromosome. This is unbalanced as there is a single copy of Xp and three copies of Xq.

47,XY,i(X)(q10)
A male showing an isochromosome of the long arm of the X chromosome in addition to a normal X and Y.

46,XX,idic(17)(p11.2)
46,XX,idic(17)(qter→p11.2::p11.2→qter)
An isodicentric chromosome composed of the long arms of chromosome 17 and the short arm material between the centromeres and the breakpoints in 17p11.2.

46,XX,i(21)(q10)
An isochromosome of the long arm of chromosome 21 has replaced one chromosome 21. There are two copies of the long arm of chromosome 21 in the isochromosome and one normal copy of chromosome 21. Even though there are effectively three copies of the long arm of chromosome 21, the normal chromosome 21 is not designated with a plus sign (+). Note that an alternative description for this same chromosomal rearrangement based on G-banding is found in Section 9.2.17.3 and makes the additional copy of chromosome 21 more obvious.

45,XX,−21,i(21)(q10)
An isochromosome of the long arm of chromosome 21 has replaced one chromosome 21. The other chromosome 21 is lost. There are two copies of the long arm of chromosome 21 in the isochromosome and no normal copies of chromosome 21.

9.2.12 Marker Chromosomes

A *marker chromosome* (**mar**) is a structurally abnormal chromosome that cannot be unambiguously identified or characterized by conventional banding cytogenetics. Numerous terms have been used in the literature to describe markers, including "supernumerary marker chromosomes (SMC)," "extra structurally abnormal chromosomes (ESAC)," "supernumerary ring chromosomes (SRC)," and "accessory chromosomes (AC)"; see Liehr et al. (2004) for a review of these abnormalities. Whenever any part of an abnormal chromosome can be recognized, it is a derivative chromosome (**der**) and can be adequately described by the nomenclature for derivative chromosomes (see Section 9.2.3). For placement of **mar** in the karyotype, see Chapter 6. In the description of a karyotype, the presence of a **mar** must be preceded by a **plus sign** (+). No attempt should be made to describe the morphology or size of markers in karyotypes. Using min mar, A-size mar, acro mar, etc., is discouraged. If such information is relevant, it should be described in words in the text.

47,XX,+mar
: One additional marker chromosome.

47,XX,t(12;16)(q13;p11.2),+mar
: One marker chromosome in addition to t(12;16).

48,X,t(X;18)(p11.2;q11.2),+2mar
: Two marker chromosomes in addition to t(X;18).

47~51,XY,t(11;22)(q24;q12),+1~5mar[cp10]
: In this tumor there are, in addition to a t(11;22), five clonally occurring markers, but not all cells contain all the markers.

When several different markers are clonally present, they may be indicated by an *Arabic* number after the symbol mar, e.g., mar1, mar2, etc. It must be stressed that this does not mean derivation of the marker from chromosome 1, chromosome 2, and so on. Multiple copies of the same marker are indicated by a multiplication sign after the mar designation, e.g., mar1×2 indicates two markers 1; mar1×3 indicates three markers 1, and so on.

48,XX,i(17)(q10),+mar1,+mar2[17]/51,XX,i(17)(q10),+mar1×3,+mar2,+mar3[13]
: There are two different markers (mar1 and mar2) in the clone with 48 chromosomes. The clone with 51 chromosomes has three copies of mar1, one copy of mar2, and in addition a third marker (mar3).

As soon as any part of an abnormal chromosome can be recognized, even if the origin of the centromere is unknown, this abnormal chromosome is referred to as a **der** and not as a mar (see Section 9.2.3).

47,XX,+der(?)t(?;15)(?;q22)
: The centromere of this abnormal chromosome is unknown and hence it is designated der(?), but part of the chromosome is composed of the chromosome 15 segment distal to band 15q22.

Double minutes, abbreviated **dmin**, represent a special kind of acentric structures that should be recorded in the karyotype when present in more than one metaphase cell. Note that the dmin should not be included in the chromosome count, and that the symbol should not be preceded by a plus sign. It is placed after any centric marker. The number of dmin per cell should be presented before the symbol either in absolute numbers or as a mean or a range.

49,XX,…,+3mar,1dmin
: A tumor with one dmin per cell.

49,XY,…,+3mar,~14dmin
: A tumor with approximately 14 dmin per cell.

49,XX,…,+3mar,9~>50dmin
: A tumor with 9 to more than 50 dmin per cell.

Acentric fragments (**ace**) other than dmin, even if present in more than one cell, should not be presented in the karyotype, but must be recorded in chromosome breakage studies (see Chapter 10).

9.2.13 Neocentromeres

A *neocentromere* is a functional centromere that has arisen or been activated within a region not known to have a centromere. A chromosome with a neocentromere may be described with the symbol **neo** or as a derivative chromosome with the assumption that a new centromere has arisen (or has been activated) within the region(s) from which the chromosome segment was derived.

47,XX,+der(3)(qter→q28:)
> An additional derivative chromosome containing segments 3q28 through 3qter. Because this segment usually does not contain a centromere, this example is a chromosome with a neocentromere. In this example, neo could be used instead of der: 47,XX,+neo(3)(qter→q28:). Also, the location of the neocentromere could be indicated by the symbol neo: 47,XX,+der(3)(qter→q28→neo→q28:). Note that the short form may not be adequate to describe this chromosome.

47,XX,der(3)(:p11→q11:),+neo(3)(pter→p11::q11→q26→neo→q26→qter)
> Chromosome 3 has been replaced by a derivative small chromosome containing the chromosome 3 centromere and by a large chromosome composed of the remaining part of chromosome 3 where a neocentromere has been activated at 3q26.

Unlike duplications in which the orientation of the duplicated segment is indicated by the order of the bands from pter to qter, a supernumerary marker chromosome containing a neocentromere may require the use of the detailed form depending on the circumstance.

47,XX,+dup(10)(qter→q25::q25→qter)
47,XX,+dup(10)(qter→q25::q25→q26→neo→q26→qter)
> A supernumerary chromosome that is a duplication of the segments between bands 10q25 and 10qter. Because this segment does not usually contain a centromere, this example is a derivative chromosome with a neocentromere. The location of the neocentromere could be indicated as shown.

9.2.14 Quadruplications

The symbol **qdp** is used. It is not possible to indicate the orientation(s) of the segment with the short form.

46,XX,qdp(1)(q23q32)
46,XX,qdp(1)(pter→q32::q23→q32::q23→q32::q23→qter)
> Quadruplication of the segment between bands 1q23 and 1q32.

9.2.15 Ring Chromosomes

A *ring*, designated by the symbol **r**, may be composed of one or several chromosomes.

Ring chromosomes derived from one chromosome

As in other rearrangements affecting a single chromosome, there is no semicolon between the band designations.

46,XX,r(7)(p15q31)
46,XX,r(7)(::p15→q31::)
    Ring chromosome in which breakage and reunion have occurred at bands 7p15 and 7q31. The segments distal to these breakpoints have been deleted.

46,XX,r(20)(p13q13.3)
46,XX,r(20)(::p13→q13.3::)
    Ring chromosome in which breakage and reunion have occurred at bands 20p13 and 20q13.3 was identified at 550 band resolution with no discernable deletion of material at either breakpoint.

Ring chromosomes derived from more than one chromosome

Ring chromosomes derived from more than one chromosome may contain one or several centromeres.

    Monocentric ring chromosomes are treated as derivative (**der**) chromosomes (see Section 9.2.3). The chromosome that provides the centromere is listed first. The orientation of the acentric segment is apparent from the order of the breakpoints.

46,XX,der(1)r(1;3)(p36.1q23;q21q27)
46,XX,der(1)(::1p36.1→1q23::3q21→3q27::)
    A ring composed of chromosome 1, with breakpoints in 1p36.1 and 1q23, and the acentric segment between bands 3q21 and 3q27 of chromosome 3.

46,XX,der(1)r(1;3)(p36.1q23;q27q21)
46,XX,der(1)(::1p36.1→1q23::3q27→3q21::)
    A ring with the same breakpoints as in the previous example, but the orientation of the acentric segment is reversed.

46,XX,der(1)r(1;?)(p36.1q23;?)
46,XX,der(1)(::1p36.1→1q23::?::)
    A ring composed of chromosome 1 with breakpoints in 1p36.1 and 1q23, and an unknown acentric segment.

If the centromere of the ring chromosome is not known, but segments from other chromosomes contained in the ring are recognized, the ring is designated der(?).

47,XX,+der(?)r(?;3;5)(?;q21q26.2;q13q33)
47,XX,+der(?)(::?→cen→?::3q21→3q26.2::5q13→5q33::)
    In this ring the origin of the centromere is unknown, but the ring contains the acentric segments 3q21 to 3q26.2 and 5q13 to 5q33.

Dicentric or *tricentric ring chromosomes* are designated by the symbol **r** preceded by the abbreviation **dic** or **trc**.

    In *dicentric ring chromosomes* (**dic r**), the sex chromosomes or the autosome with the lowest number is specified first.

47,XX,+dic r(1;3)(p36.1q32;p24q26.2)
47,XX,+dic r(1;3)(::1p36.1→1q32::3p24→3q26.2::)
    A dicentric ring composed of chromosomes 1 and 3 in which 1q32 is fused with 3p24 and 3q26.2 is fused with 1p36.1.

In *tricentric ring chromosomes* (**trc r**), the sex chromosomes or the autosome with the lowest number is specified first. The orientation of the chromosomes will be apparent from the order of the breakpoints.

47,XX,+trc r(1;3;12)(p36.1q32;q26.3p24;p12q23)
47,XX,+trc r(1;3;12)(::1p36.1→1q32::3q26.3→3p24::12p12→12q23::)
    A tricentric ring in which 1q32 is fused with 3q26.3, 3p24 with 12p12, and 12q23 with 1p36.1.

When the origin of the ring is known, the description of the ring is placed in the appropriate chromosome number order.

49,XX,+1,+3,r(7),+8

When the origin of the ring is not known, the presence of the ring, preceded by a plus sign (+), is indicated at the end of the karyotype, but before any other marker chromosome (see Chapter 6).

50,XX,+1,+3,+8,+r
51,XY,+1,+3,+8,+r,+mar

Different rings may be indicated by an *Arabic* number after the symbol r, e.g., r1, r2, etc., whereas several copies of unidentified rings are indicated by the appropriate number before the ring symbol, e.g., 5r.

53,XX,…,+r1,+r2
    Two distinctly different clonally occurring rings. Note that the ring designations r1 and r2 do not mean derivation from chromosomes 1 and 2. When the origin of a ring is known, the appropriate chromosome is placed in parentheses, e.g., r(1), r(2), etc.

53,XY,…,+5r
    A total of five rings but it is not known if any of the rings are identical.

9.2.16 Telomeric Associations

The symbol **tas** is used to describe a *telomeric association*, which is typically a single cell abnormality and therefore not reported as a clonal abnormality. In telomeric associations between two chromosomes, the sex chromosome or the autosome with the lowest number is specified first. When more than two chromosomes are involved, the "end" chromosome which has the lowest number, or is one of the sex chromosomes, is specified first, followed by the other chromosomes in the order they are associated with the chromosome listed first. The terminal bands of the chromosomes involved in telomeric association(s) are given in the second parentheses; the orientation of the chromosomes will be apparent from the order in which the bands are listed. Chromosomes involved in telomeric associations are counted as separate chromosomes, as it is not proven that there is true chromosome fusion with terminal breakpoints.

46,XX,tas(12;13)(q24.3;q34)
46,XX,tas(12;13)(12pter→12qter→13qter→13pter)
    Association between the telomeric regions of the long arms of chromosomes 12 and 13.

46,Y,tas(X;12;3)(q28;p13q24.3;q29)
46,Y,tas(X;12;3)(Xpter→Xqter→12pter→12qter→3qter→3pter)
   Association between the telomeric regions of Xq and 12p, and 12q and 3q.

46,X,tas(1;X;12;7)(p36.3;q28p22.3;p13q24.3;p22)
46,X,tas(1;X;12;7)(1qter→1pter→Xqter→Xpter→12pter→12qter→7pter→7qter)
   Association between the telomeric regions of 1p and Xq, Xp and 12p, and 12q and 7p.

9.2.17 Translocations

9.2.17.1 Reciprocal Translocations

The symbol **t** is used to denote a *translocation*, which refers to the exchange of terminal segments of chromosomes. When a translocation is balanced, there is no discernable gain or loss of chromosomal material. In translocations affecting two chromosomes, the sex chromosome or the autosome with the lowest number is always specified first. The same rule is followed in translocations involving three chromosomes, but in these rearrangements the chromosome specified next is the one receiving a segment from the one listed first, and the chromosome specified last is the one donating a segment to the first chromosome listed. Whenever applicable, the same rules should be followed in four-break and more complex balanced translocations. In order to distinguish homologous chromosomes, one of the numerals may be underlined (single underlining).

Two-break rearrangements

46,XY,t(2;5)(q21;q31)
46,XY,t(2;5)(2pter→2q21::5q31→5qter;5pter→5q31::2q21→2qter)
   Breakage and reunion have occurred at bands 2q21 and 5q31. The segments distal to these bands have been exchanged.

46,XY,t(2;5)(p12;q31)
46,XY,t(2;5)(5qter→5q31::2p12→2qter;5pter→5q31::2p12→2pter)
   Breakage and reunion have occurred at bands 2p12 and 5q31. The segments distal to these bands have been exchanged.

46,X,t(X;13)(q27;q12)
46,X,t(X;13)(Xpter→Xq27::13q12→13qter;13pter→13q12::Xq27→Xqter)
   Breakage and reunion have occurred at bands Xq27 and 13q12. The segments distal to these bands have been exchanged. Since one of the chromosomes involved in the translocation is a sex chromosome, it is designated first. Note that the correct designation is 46,X,t(X;13) and not 46,XX,t(X;13). Similarly, an identical translocation in a male should be designated 46,Y,t(X;13) and not 46,XY,t(X;13).

46,t(X;Y)(q22;q11.2)
46,t(X;Y)(Xpter→Xq22::Yq11.2→Yqter;Ypter→Yq11.2::Xq22→Xqter)
   A reciprocal translocation between an X chromosome and a Y chromosome with breakpoints at bands Xq22 and Yq11.2.

46,t(X;18)(p11.2;q11.2),t(Y;1)(q11.2;p31)
46,t(X;18)(18qter→18q11.2::Xp11.2→Xqter;18pter→18q11.2::Xp11.2→Xpter),t(Y;1)
(Ypter→Yq11.2::1p31→1pter;Yqter→Yq11.2::1p31→1qter)

> Two reciprocal translocations, each involving one sex chromosome. Breakage and reunion have occurred at bands Xp11.2 and 18q11.2 as well as at bands Yq11.2 and 1p31. Abnormalities of the X chromosome are listed before those of the Y chromosome.

Three-break rearrangements

46,XX,t(2;7;5)(p21;q22;q23)
46,XX,t(2;7;5)(5qter→5q23::2p21→2qter;7pter→7q22::2p21→2pter;5pter→
5q23::7q22→7qter)

> The segment on chromosome 2 distal to 2p21 has been translocated onto chromosome 7 at band 7q22, the segment on chromosome 7 distal to 7q22 has been translocated onto chromosome 5 at 5q23, and the segment of chromosome 5 distal to 5q23 has been translocated onto chromosome 2 at 2p21.

46,X,t(X;22;1)(q24;q11.2;p33)
46,X,t(X;22;1)(Xpter→Xq24::1p33→1pter;22pter→22q11.2::Xq24→Xqter;22qter→
22q11.2::1p33→1qter)

> The segment on one X chromosome distal to Xq24 has been translocated onto chromosome 22 at band 22q11.2, the segment distal to 22q11.2 has been translocated onto chromosome 1 at band 1p33, and the segment distal to 1p33 has been translocated onto the X chromosome at band Xq24.

46,XX,t(2;7;<u>7</u>)(q21;q22;p13)
46,XX,t(2;7;<u>7</u>)(2pter→2q21::<u>7</u>p13→<u>7</u>pter;7pter→7q22::2q21→2qter;7qter→7q22::<u>7</u>p13→
<u>7</u>qter)

> The segment on chromosome 2 distal to 2q21 has been translocated onto chromosome 7 at band 7q22, the segment on chromosome 7 distal to 7q22 has been translocated onto the homologous chromosome 7 at band 7p13, and the segment distal to 7p13 on the latter chromosome has been translocated onto chromosome 2 at 2q21. Underlining is used only to emphasize that the chromosomes are homologous. However, this is usually not necessary since if the same chromosome 7 had been involved, the resulting chromosome 7 would have to be described as a derivative chromosome.

Four-break and more complex rearrangements

46,XX,t(3;9;22;21)(p13;q34;q11.2;q21)
46,XX,t(3;9;22;21)(21qter→21q21::3p13→3qter;9pter→9q34::3p13→3pter;22pter→
22q11.2::9q34→9qter;21pter→21q21::22q11.2→22qter)

> The segment of chromosome 3 distal to 3p13 has been translocated onto chromosome 9 at 9q34, the segment of chromosome 9 distal to 9q34 has been translocated onto chromosome 22 at 22q11.2, the segment of chromosome 22 distal to 22q11.2 has been translocated onto chromosome 21 at 21q21, and the segment of chromosome 21 distal to 21q21 has been translocated onto chromosome 3 at 3p13.

46,XX,t(3;9;<u>9</u>;22)(p13;q22;q34;q11.2)
46,XX,t(3;9;<u>9</u>;22)(22qter→22q11.2::3p13→3qter;9pter→9q22::3p13→3pter;<u>9</u>pter→
<u>9</u>q34::9q22→9qter;22pter→22q11.2::<u>9</u>q34→<u>9</u>qter)

> Four-break rearrangement involving the two homologous chromosomes 9. The segment on chromosome 3 distal to 3p13 has been translocated onto chromosome 9 at band 9q22, the segment on chromosome 9 distal to 9q22 has been translocated onto the homologous chromosome 9 at 9q34, the segment on the latter chromosome 9 distal to 9q34 has been translocated onto chromosome 22 at 22q11.2, and the segment on chromosome 22 distal to 22q11.2 has been translocated onto chromosome 3 at 3p13.

46,XY,t(5;6)(q13q23;q15q23)
46,XY,t(5;6)(5pter→5q13::6q15→6q23::5q23→5qter;6pter→6q15::5q13→5q23::6q23→6qter)
: Four-break rearrangement involving two chromosomes. The segment between bands 5q13 and 5q23 in chromosome 5 and the segment between bands 6q15 and 6q23 in chromosome 6 have been exchanged.

46,XX,t(5;14;9)(q13q23;q24q21;p12p23)
46,XX,t(5;14;9)(5pter→5q13::9p12→9p23::5q23→5qter;14pter→14q21::5q13→5q23::14q24→14qter;9pter→9p23::14q21→14q24::9p12→9qter)
: Reciprocal six-break translocation of three interstitial segments. The segment between bands 5q13 and 5q23 on chromosome 5 has replaced the segment between bands 14q21 and 14q24 on chromosome 14, the segment 14q21q24 has replaced the segment between bands 9p23 and 9p12 on chromosome 9, and the segment 9p23p12 has replaced the segment 5q13q23. The orientations of the segments in relation to the centromere are apparent from the order of the bands. The segments 9p23p12 and 14q21q24 are inverted.

The *derivative chromosomes* produced by malsegregation of a reciprocal translocation should be described using the conventions outlined in Section 9.2.3.

9.2.17.2 Whole-Arm Translocations

Whole-arm translocations are described by assigning the breakpoints to the centromeric bands p10 and q10 according to the morphology of the abnormal chromosomes. In *balanced whole-arm exchanges*, the breakpoint in the chromosome which has the lowest number, or is a sex chromosome, is assigned to p10.

46,XY,t(1;3)(p10;q10)
46,XY,t(1;3)(1pter→1p10::3q10→3qter;3pter→3p10::1q10→1qter)
: Reciprocal whole-arm translocation in which the short arm of chromosome 1 has been fused at the centromere with the long arm of chromosome 3 and the long arm of chromosome 1 has been fused with the short arm of chromosome 3.

46,XY,t(1;3)(p10;p10)
46,XY,t(1;3)(1pter→1p10::3p10→3pter;1qter→1q10::3q10→3qter)
: Reciprocal whole-arm translocation in which the short arms of chromosomes 1 and 3 and the long arms of these chromosomes, respectively, have been fused at the centromeres.

In the description of karyotypes containing derivative chromosomes resulting from *unbalanced whole-arm translocations* (see Section 9.2.3), the derivative chromosome (**der**) replaces the two normal chromosomes involved in the translocation. The two missing normal chromosomes are not specified. The imbalance(s) can be inferred from the karyotype designation.

45,XX,der(1;3)(p10;q10)
45,XX,der(1;3)(1pter→1p10::3q10→3qter)
: A derivative chromosome consisting of the short arm of chromosome 1 and the long arm of chromosome 3. The missing chromosomes 1 and 3 are not indicated since they are replaced by the derivative chromosome. The karyotype contains one normal chromosome 1, one normal chromosome 3, and the der(1;3). The resulting net imbalance of this abnormality is monosomy for the long arm of chromosome 1 and monosomy for the short arm of chromosome 3.

46,XX,+1,der(1;3)(p10;q10)
  A derivative chromosome consisting of the short arm of chromosome 1 and the long arm of chromosome 3 (same as above) has replaced one chromosome 1 and one chromosome 3. There are, however, two normal chromosomes 1, i.e., an additional chromosome 1 in relation to the expected loss due to the der(1;3). Consequently, this gain is indicated as +1. The karyotype contains two normal chromosomes 1, one normal chromosome 3, and the der(1;3). The resulting net imbalance is trisomy for 1p and monosomy for 3p.

46,XX,der(1;3)(p10;q10),+3
  A derivative chromosome consisting of the short arm of chromosome 1 and the long arm of chromosome 3 (same as above) has replaced one chromosome 1 and one chromosome 3. There are, however, two normal chromosomes 3, i.e., an additional chromosome 3 in relation to the expected loss due to the der(1;3). Consequently, this gain is indicated as +3. The karyotype contains one normal chromosome 1, two normal chromosomes 3, and the der(1;3). The resulting net imbalance is monosomy for 1q and trisomy for 3q.

47,XX,+der(1;3)(p10;q10)
  An extra derivative chromosome consisting of the short arm of chromosome 1 and the long arm of chromosome 3 (same as above). There are two normal chromosomes 1, two normal chromosomes 3, and the der(1;3). The resulting net imbalance is trisomy for 1p and trisomy for 3q.

44,XY,–1,der(1;3)(p10;q10)
  A derivative chromosome consisting of the short arm of chromosome 1 and the long arm of chromosome 3 (same as above) has replaced one chromosome 1 and one chromosome 3. There is, however, no normal chromosome 1, indicated as –1 in relation to the expected presence of one chromosome 1 due to the der(1;3). The karyotype contains no chromosome 1, one normal chromosome 3, and the der(1;3). The resulting net imbalance is nullisomy for 1q, monosomy for 1p, and monosomy for 3p.

9.2.17.3 Robertsonian Translocations

These special types of translocations originate through translocation of the acrocentric chromosomes 13–15 and 21–22. The breakpoints mostly occur in the short arms, resulting in dicentric chromosomes. Breaks may also occur in one short arm and one long arm of the participating chromosomes, resulting in monocentric rearrangements. Usually there is simultaneous loss of the remaining short arms. Although either **rob** or **der** can adequately describe these whole-arm translocations, **der** is the preferred designation. The abbreviation rob should never be used in the description of acquired abnormalities.

45,XX,der(13;21)(q10;q10)
  Breakage and reunion have occurred at band 13q10 and band 21q10 in the centromeres of chromosomes 13 and 21. The derivative chromosome has replaced one chromosome 13 and one chromosome 21 and there is no need to indicate the missing chromosomes. The karyotype contains one normal chromosome 13, one normal chromosome 21, and the der(13; 21). The resulting net imbalance is loss of the short arms of chromosomes 13 and 21.

46,XX,der(13;21)(q10;q10),+21
  A derivative chromosome consisting of the long arm of chromosome 13 and the long arm of chromosome 21 (same as above) has replaced one chromosome 13 and one chromosome 21. There are, however, two normal chromosomes 21, i.e., an additional chromosome 21 in relation to the expected loss due to the der(13;21). Consequently, this gain is indicated as +21. The karyotype contains one normal chromosome 13, two normal chromosomes 21, and the der(13;21). The resulting net imbalance is loss of the short arm of chromosome 13 and trisomy for the long arm of chromosome 21.

46,XX,+13,der(13;21)(q10;q10)
A derivative chromosome consisting of the long arm of chromosome 13 and the long arm of chromosome 21 (same as above) has replaced one chromosome 13 and one chromosome 21. The karyotype contains two normal chromosomes 13, one normal chromosome 21, and the der(13;21). Consequently, this gain is indicated as +13. The resulting net imbalance is loss of the short arm of chromosome 21 and trisomy for the long arm of chromosome 13.

If only a single chromosome is involved in the rearrangement, the extra chromosome is indicated by the 46 count in the presence of a whole-arm rearrangement and the addition of a normal chromosome.

46,XX,+21,der(21;21)(q10;q10)
A derivative chromosome composed of the long arms of chromosome 21. There are two copies of the long arm of chromosome 21 in the derivative chromosome and one normal chromosome 21, indicated by the 46 count. The normal chromosome 21 is designated with a (+) sign. See also Section 9.2.11 for an alternate description of this same rearrangement.

If it is proven that the derivative chromosome resulting from a whole-arm translocation is *dicentric*, i.e., the breakpoints have been assigned to p11.2 or q11.2, the abbreviation **dic** should be used and the dicentric chromosome should be described accordingly (see Section 9.2.4).

9.2.17.4 Jumping Translocations

These can be described by the standard nomenclature for translocations. The clones are presented in the same order as unrelated clones, i.e., in order of decreasing frequency (see Section 11.1.6).

46,XX,t(4;7)(q35;q11.2)[6]/46,XX,t(1;7)(p36.3;q11.2)[4]/46,XX,t(7;9)(q11.2;p24)[3]
Three clonal translocations involving band 7q11.2. The segment 7q11.2qter is translocated to bands 1p36.3, 4q35, and 9p24.

9.2.18 Tricentric Chromosomes

The symbol **trc** is used. The "end" chromosome which has the lowest number, or is one of the sex chromosomes, is specified first. The other chromosomes are listed in the order they are attached to the chromosome listed first. The orientation of the chromosomes will be apparent from the order of the breakpoints specified in the second parentheses. A tricentric chromosome is counted as one chromosome.

44,XX,trc(4;12;9)(q31.2;q22p13;q34)
44,XX,trc(4;12;9)(4pter→4q31.2::12q22→12p13::9q34→9pter)
A tricentric chromosome in which band 4q31.2 is fused with 12q22 and 12p13 is fused with 9q34.

9.2.19 Triplications

The symbol **trp** is used. It is not possible to indicate the orientation(s) of the segment in the short form, but this can be done with the detailed form.

46,XX,trp(1)(q21q32)
46,XX,trp(1)(pter→q32::q21→q32::q21→qter)
   Triplication of the segment between bands 1q21 and 1q32, one of several possible orientations of the triplications of this segment.

46,XX,trp(1)(q32q21)
46,XX,trp(1)(pter→q32::q32→q21::q21→qter)
   Triplication of the segment between bands 1q21 and 1q32 in an opposite orientation to the above example.

9.3 Multiple Copies of Rearranged Chromosomes

The **multiplication sign** (×) can be used to describe two or more copies of a structurally rearranged chromosome. The number of copies (×2, ×3, etc.) should be placed after the abnormality. The multiplication sign should not be used to denote multiple copies of normal chromosomes.

46,XX,del(6)(q13q23)×2
   Two deleted chromosomes 6 with breakpoints at bands 6q13 and 6q23, and no normal chromosome 6. Since the two abnormal chromosomes replace the two normal chromosomes, there is no need to indicate the missing normal chromosomes.

48,XY,+del(6)(q13q23)×2
   Two normal chromosomes 6 plus two additional deleted chromosomes 6 with breakpoints at bands 6q13 and 6q23.

47,XX,del(6)(q13q23)×2,+del(6)(q13q23)
   There are three copies of a deleted chromosome 6 and no normal chromosome 6, i.e., two of the deleted chromosomes replace the two normal chromosomes 6. Note that the supernumerary deleted chromosome 6 has to be preceded by a plus sign.

48,XX,del(6)(q13q23)×2,+7,+7
   Two deleted chromosomes 6 replace the two normal chromosomes 6; in addition, there are two extra chromosomes 7.

48,XX,t(8;14)(q24.1;q32),+der(14)t(8;14)×2
   A balanced t(8;14) plus two additional copies of the derivative chromosome 14.

92,XXXX,t(8;14)(q24.1;q32)×2
   A tetraploid clone with two balanced t(8;14). The two derivative chromosomes 8 and 14 replace two normal chromosomes 8 and 14. There are two normal chromosomes 8 and 14.

94,XXYY,t(8;14)(q24.1;q32)×2,+der(14)t(8;14)×2
   A hypertetraploid clone with two balanced t(8;14) plus two additional copies of the derivative chromosome 14. There are two normal chromosomes 8 and 14.

93,XXXX,t(8;14)(q24.1;q32)×2,der(14)t(8;14)×2,+der(14)t(8;14)
A hypertetraploid clone with two balanced t(8;14) and three extra copies of the derivative chromosome 14, i.e., there are in total five der(14), four of which replace the normal chromosomes 14; consequently there is no normal chromosome 14.

94,XXYY,t(8;14)(q24.1;q32)×2,+14,der(14)t(8;14)×2,+der(14)t(8;14)
A hypertetraploid clone with two balanced t(8;14), three extra copies of the derivative chromosome 14, and one normal chromosome 14.

47,XX,+8,i(8)(q10)×2
47,XX,i(8)(q10),+i(8)(q10)
Alternative descriptions of the same chromosome complement with one normal chromosome 8 and two copies of an isochromosome for the long arm of chromosome 8.

10 Chromosome Breakage

This section provides a nomenclature for the chromatid and chromosome aberrations that may be observed in, for example, constitutional chromosome breakage syndromes or following clastogenic exposure. Since many aberrations of this kind are scored on unbanded material, recommendations are given first for non-banded preparations and then for banded preparations.

10.1 Chromatid Aberrations

10.1.1 Non-Banded Preparations

A *chromatid* (**cht**) aberration involves only one chromatid in a chromosome at a given locus.

A *chromatid gap* (**chtg**) is a non-staining region (achromatic lesion) of a single chromatid in which there is minimal misalignment of the chromatid.

A *chromatid break* (**chtb**) is a discontinuity of a single chromatid in which there is a clear misalignment of one of the chromatids.

A *chromatid exchange* (**chte**) is the result of two or more chromatid lesions and the subsequent rearrangement of chromatid material. Exchanges may be between chromatids of different chromosomes (*interchanges*) or between or within chromatids of one chromosome (*intrachanges*). In the case of interchanges, it will generally be sufficient to indicate whether the configuration is *triradial* (**tr**) when there are three arms to the pattern, *quadriradial* (**qr**) when there are four, or *complex* (**cx**) when there are more than four. The number of centromeres may be indicated within parentheses (1 cen, 2 cen, etc.). When necessary, exchanges may be classified in more detail. *Asymmetrical* exchanges inevitably result in the formation of an acentric fragment, whereas *symmetrical* ones do not. In *complete exchanges* all the broken ends are rejoined, but not in *incomplete* ones. In asymmetrical exchanges, the incompleteness may be proximal when the broken ends nearest the centromere are not rejoined or distal when the ends farthest from the centromere are not rejoined. Intra-arm events include duplications, deletions, paracentric inversions, and isochromatid breaks showing sister reunion. It should be noted that these terms are only descriptive and do not imply knowledge of the origin of the aberrations.

Sister chromatid exchange, detectable only by special staining methods, results from the interchange of homologous segments between two chromatids of one chromosome. The symbol **sce** can be used to describe this event.

10.1.2 Banded Preparations

Some chromatid aberrations can be defined more precisely or can be recognized with certainty only in banded preparations; e.g., a *chromatid deletion* (**cht del**) is the absence of a banded sequence from only one of the two chromatids of a single chromosome. A *chromatid inversion* (**cht inv**) is the reversal of a banded sequence of only one of the two chromatids of a single chromosome. Both are subclasses of *chromatid exchanges* (**chte**).

Where it is desired to specify the location of a chromatid aberration, the appropriate symbol can be followed by the band designation.

chtg(4)(q25)	Chromatid gap in chromosome 4 at band 4q25.
chtb(4)(q25)	Chromatid break in chromosome 4 at band 4q25.
chte(4;10)(q25;q22)	Chromatid exchange involving chromosomes 4 and 10 at bands 4q25 and 10q22.
cht del(1)(q12q25)	Chromatid deletion in chromosome 1 with loss of the segment between bands 1q12 and 1q25.
cht inv(1)(q12q25)	Chromatid inversion in chromosome 1 with reversal of the segment between bands 1q12 and 1q25.
sce(4)(q25q33)	Sister chromatid exchanges in chromosome 4 at bands 4q25 and 4q33.

10.2 Chromosome Aberrations

10.2.1 Non-Banded Preparations

A *chromosome* (**chr**) aberration involves both chromatids of a single chromosome at the same locus.

A *chromosome gap* (**chrg**) is a non-staining region (achromatic lesion) at the same locus in both chromatids of a single chromosome in which there is minimal misalignment of the chromatids. The term *chromosome gap* is synonymous with *isolocus gap* and *isochromatid gap*.

A *chromosome break* (**chrb**) is a discontinuity at the same locus in both chromatids of a single chromosome, giving rise to an acentric fragment and an abnormal monocentric chromosome. This fragment is therefore a particular type of acentric fragment (**ace**), and chrb should be used only when the morphology indicates that the fragment is the result of a single event. The term *chromosome break* is synonymous with *isolocus break* and *isochromatid break*.

A *chromosome exchange* (**chre**) is the result of two or more chromosome lesions and the subsequent relocation of both chromatids of a single chromosome to a new position on the same or on another chromosome. It may be symmetrical (e.g., reciprocal translocation) or asymmetrical (e.g., dicentric formation).

A *minute* (**min**) is an acentric fragment smaller than the width of a single chromatid. It may be single or double. In the special situation when *double minutes* are present clonally in tumor cells, the symbol **dmin** is used; see Section 9.2.12.

Pulverization (**pvz**) indicates a situation where a cell contains both chromatid and/or chromosome gaps and breaks which are not normally associated with exchanges and are present in such numbers that they cannot be enumerated. Occasionally, one or more chromosomes in a cell are pulverized while the remaining chromosomes are of normal morphology; e.g., pvz(1) is a pulverized chromosome 1. The term *chromothripsis* (**cth**) describes complex patterns of alternating copy number changes (commonly alternating disomy and heterozygous loss) along the length of a chromosome or chromosome segment (Stephens et al., 2011). As these complex rearrangements cannot be visualized on banded or non-banded chromosomes, this symbol is used after microarray analysis (see Section 14.2.7).

Premature chromosome condensation (**pcc**) occurs when an interphase nucleus is prematurely induced to enter mitosis. A pcc may involve a G1 or a G2 nucleus. The chromatin of S-phase nuclei undergoing pcc often appears to be pulverized.

The term *premature centromere division* (**pcd**) may be used to describe premature separation of centromeres in metaphase. The pcd may affect one or more chromosomes in a fraction of the cells.

A *marker chromosome* (**mar**) is a structurally rearranged chromosome in which no part can be identified (see Section 9.2.12).

10.2.2 Banded Preparations

When banded preparations allow adequate identification of chromosome segments or chromosome aberrations, the nomenclature system recommended throughout this book can be used. When not, the observations can be described in words.

10.3 Scoring of Aberrations

In the scoring of aberrations, the main types are **chtg**, **chtb**, **chte**, **chrg**, **chrb**, **ace**, **min**, **r**, **dic**, **tr**, **qr**, **der**, and **mar**, and reports should, where possible, give the data under these headings. It is recognized, however, that aberrations are frequently grouped to give adequate numbers for statistical analysis or for some other reason. When this is done, it should be indicated how the groupings relate to the aberrations listed above, e.g.:

chromatid aberrations	chtg, chtb, chte
fragments (= deletions)	chrb, ace, min
asymmetric aberrations	ace, dic, r

The data should not be presented as deduced breakages per cell but in such a manner that it is possible to calculate the number of aberrations per cell.

11 Neoplasia

Described below are definitions of terms and recommendations related to abnormalities commonly seen in neoplasia.

11.1 Clones and Clonal Evolution

11.1.1 Definition of a Clone

A clone is defined as a cell population derived from a single progenitor. It is common practice to infer a clonal origin when a number of cells have the same or closely related abnormal chromosome complements. A clone is therefore not necessarily completely homogeneous because subclones may have evolved during the development of the tumor. A clone must have at least two cells with the same aberration if the aberration is a chromosome gain or a structural rearrangement. If the abnormality is loss of a chromosome, the same loss must be present in at least three cells to be accepted as clonal. The term may need to be operationally defined because the criteria for acceptance will depend on, e.g., the number of cells examined, the nature of the aberration involved, the type of culture, and the time cells spend *in vitro* prior to harvest. In the special situation when *in situ* preparations are analyzed, the same structural rearrangement or chromosomal gain must be present in at least two metaphase cells from either different primary culture slides, or from well-separated areas or different cell colonies on the same slide. Loss of a single chromosome must be detected in at least three such cells. However, two cells with identical losses of one or more chromosomes and the same chromosome gain or structural aberration(s) may be considered clonal and included in the nomenclature.

46,XY,del(5)(q13q33),–7,+8[2]/46,XY[18]
46,XX,t(6;11)(q27;q23)[16]/45,X,–X,t(6;11)(q27;q23)[2]/46,XX[1]

The karyotype designations of different clones and subclones are separated by a **slant line** (/). For order of clone presentation, see Sections 11.1.4 and 11.1.6.

The general rule in tumor cytogenetics is that only the clonal chromosomal abnormalities found in a tumor should be reported. If, for special reasons, nonclonal aberrations are presented, then these must be clearly separated from the clonal abnormalities and should not be part of the description of the tumor karyotype. When the same abnormal clone has been found in an initial and follow-up study, even in a single cell, it should be reported in the karyotype.

46,XX,t(9;22)(q34;q11.2)[1]/46,XX[19]

Similarly, when a single abnormal cell is confirmed by a different method (e.g., FISH), and thus shown to be clonal, it should be reported in the karyotype (see Section 13.3).

46,XX,del(20)(q11.2q13.3)[1]/46,XX[19].nuc ish(D20S108×1)[40/200]

When additional abnormalities are seen in a single cell, but not proven to be present with another method, they should not be listed in the nomenclature but, if appropriate, can be discussed in the interpretation.

11.1.2 Clone Size

The number of cells that constitute a clone is given in **square brackets** [ ] after the karyotype. When all cells are normal, the number of cells is still specified. Cytogenetically related clones are written in order of increasing complexity, irrespective of the size of the clone.

46,XX[20]
    A normal female karyotype identified in 20 metaphase cells.

46,XX,t(8;21)(q22;q22)[23]
    A clone with t(8;21) identified in 23 metaphase cells.

46,XX,t(9;22)(q34;q11.2)[18]/45,XX,der(7;9)(q10;q10)t(9;22),der(22)t(9;22)[2]
    A clone with 46 chromosomes identified with a t(9;22) as the sole aberration in 18 cells. A subclone with 45 chromosomes was identified in 2 cells with the derivative chromosomes 9 from the t(9;22) further involved in a whole-arm translocation with chromosome 7.

46,XY,t(8;21)(q22;q22)[26]/47,XY,t(8;21),+21[7]
46,XY,t(8;21)(q22;q22)[26]/47,idem,+21[7]
46,XY,t(8;21)(q22;q22)[26]/47,sl,+21[7]
    A clone with 46 chromosomes identified with a t(8;21) as the sole aberration in 26 cells and a subclone with 47 chromosomes with the t(8;21) and trisomy 21 in 7 cells. Alternatively, the terms idem or sl may be used to describe subclones (see Section 11.1.4 for details); however, the terms idem and sl must never be intermixed when describing a single tumor sample.

11.1.3 Mainline

The *mainline* is the most frequent chromosome constitution of a tumor cell population. It is a purely quantitative term to describe the largest clone, and does not necessarily indicate the most basic one in terms of progression. In some situations, when two or more clones are of exactly the same size, a tumor may have more than one mainline. The terms **idem** and **sl** may be used to describe subclones (see Section 11.1.4 for details).

46,XX,t(9;22)(q34;q11.2)[3]/47,XX,+8,t(9;22)[17]
46,XX,t(9;22)(q34;q11.2)[3]/47,sl,+8[17]
46,XX,t(9;22)(q34;q11.2)[3]/47,idem,+8[17]
> The clone with 47 chromosomes represents the mainline, although it has most probably evolved from the clone with 46 chromosomes.

46,XX,der(2)t(2;5)(p23;q35)[10]/47,XX,+2,der(2)t(2;5)[16]
46,XX,der(2)t(2;5)(p23;q35)[10]/47,sl,+2[16]
46,XX,der(2)t(2;5)(p23;q35)[10]/47,idem,+2[16]
> The clone with 47 chromosomes represents the mainline, although it has most probably evolved from the clone with 46 chromosomes.

11.1.4 Stemline, Sideline and Clonal Evolution

Cytogenetically related clones (subclones) are presented, as far as possible, in order of increasing complexity, irrespective of the size of the clone. The s*temline* (**sl**) is the most basic clone of a tumor cell population and is listed first. All additional deviating subclones are termed *sidelines* (**sdl**). To describe the stemlines or sidelines, these symbols, or the term **idem** (Latin = same), can be used. If more than one sideline is present, these may be referred to as sdl1, sdl2, and so on.

In tumors with subclones the term **idem** can be used, followed by the additional changes in relation to the stemline, which is listed first. Note that idem always refers to the karyotype listed first. This means that in tumors with multiple subclones each clonal change in addition to the first karyotype will have to be repeated. It also means that all plus and minus signs only refer to changes in relation to the stemline karyotype. As an alternative, for more than one sideline, sl and sdl could be used. Note that when two or more stemlines are present, there may also exist two or more sdl1, sdl2 and so on, which will reduce clarity. **In such instances idem is preferred.**

46,XX,t(9;22)(q34;q11.2)[3]/47,sl,+8[17]/48,sdl1,+9[3]/49,sdl2,+11[12]
46,XX,t(9;22)(q34;q11.2)[3]/47,idem,+8[17]/48,idem,+8,+9[3]/49,idem,+8,+9,+11[12]
> The clone with 46 chromosomes represents the stemline; the three subclones with 47, 48 and 49 chromosomes are sidelines. In the subclone with 47 chromosomes, the designation sl indicates the presence of the abnormal chromosomes seen in the stemline, i.e., t(9;22)(q34;q11.2) in addition to +8; this subclone is sideline 1 (sdl1). In the subclone with 48 chromosomes (sdl2), the designation sdl1 indicates the presence of the abnormal chromosomes seen in the first sideline, i.e., t(9;22)(q34;q11.2),+8 in addition to +9, and so on. As an alternative, in each subclone the translocation 9;22 is replaced by idem.

46,XX,t(8;21)(q22;q22)[12]/45,sl,–X[19]/46,sdl1,+8[5]/47,sdl2,+9[8]
46,XX,t(8;21)(q22;q22)[12]/45,idem,–X[19]/46,idem,–X,+8[5]/47,idem,–X,+8,+9[8]
> The clone with t(8;21) as the sole anomaly is the most basic one and hence represents the stemline; the other subclones are listed in order of increasing karyotypic complexity of the aberrations acquired during clonal evolution.

48,XX,t(12;16)(q13;p11.1),…[23]/49,sl,+6[8]/50,sdl,+7,–8,+9[4]
48,XX,t(12;16)(q13;p11.1),…[23]/49,idem,+6[8]/50,idem,+6,+7,–8,+9[4]
> The subclone with 49 chromosomes has all the abnormalities seen in the stemline plus an extra chromosome 6; the subclone with 50 chromosomes has all sdl abnormalities in addition to trisomy 7, monosomy 8, and trisomy 9.

The term **sl** or **sdl times a number** (×2, ×3, etc.) may be used to designate aberrant polyploid clones. Alternatively, the term **idem times a number** (×2, ×3, etc.) may be used to designate aberrant polyploid clones. Additional abnormalities in the polyploid clone may then be indicated using conventional terminology (see Sections 8.1 and 9.1).

26,X,+4,+6,+21[3]/52,idem×2[13]
> A near-haploid clone with two copies of chromosomes 4, 6, and 21, and a single copy of all other chromosomes is the stemline. An abnormal subclone, a doubling of the near-haploid clone (likely due to endoreduplication), is also identified.

46,XY,t(9;22)(q34;q11.2)[3]/92,sl×2[5]/93,sdl,+8[2]
46,XY,t(9;22)(q34;q11.2)[3]/92,idem×2[5]/93,idem×2,+8[2]
> The clone with the t(9;22) is the stemline. Two additional abnormal subclones are identified, one a doubling product or tetraploid subclone of the stemline (sl) and a near-tetraploid (sdl) subclone with gain of chromosome 8. As an alternative, idem may be used, but all subclones refer back to the stemline.

46,XY,t(9;22)(q34;q11.2)[3]/47,sl,+8[10]/48,sdl,+der(22)t(9;22)[4]/47,sl,+19[3]
46,XY,t(9;22)(q34;q11.2)[3]/47,idem,+8[10]/48,idem,+8,+der(22)t(9;22)[4]/47,idem,+19[3]
> The clone with the t(9;22) as the sole abnormality is the stemline. Three additional abnormal subclones are identified, one with a trisomy 8 from the stemline (sl) (now termed sdl), one that has an additional derivative chromosome 22 in the previous clone or sideline (sdl), and one with trisomy 19 of the stemline.

45,XY,–7[5]/46,sl,+8[6]/46,XY,t(9;22)(q34;q11.2)[3]/92,sl2×2[5]/93,sl2×2,+8[2]
45,XY,–7[5]/46,idem,+8[6]/46,XY,t(9;22)(q34;q11.2)[3]/92,XXYY,t(9;22)×2[5]/93,XXYY,t(9;22)×2,+8[2]
> In tumors with unrelated clones, there may be clonal evolution arising from each unrelated clone. In this instance, the first stemline shows monosomy 7 and is designated sl in the subclone showing trisomy 8. The second stemline shows t(9;22) and is designated sl2 in the subclone showing tetraploidy. Further clonal evolution is found in a sideline showing gain of chromosome 8, but to avoid confusion between sidelines of sl1 and sl2, the use of the term sdl is avoided when referring to a second stemline. The alternative use of idem is listed below for comparison.

48,XX,t(12;16)(q13;p11.1),+17,+20[31]/96,sl×2[6]
48,XX,t(12;16)(q13;p11.1),+17,+20[31]/96,idem×2[6]
> The subclone with 96 chromosomes represents a doubling product of the stemline with 48 chromosomes.

48,XX,t(12;16)(q13;p11.1),+17,+20[27]/97,sl×2,+8[3]
48,XY,t(12;16)(q13;p11.1),+17,+20[27]/97,idem×2,+8[3]
> The subclone with 97 chromosomes represents a doubling product of the hyperdiploid stemline and also has an extra chromosome 8, i.e., there are five chromosomes 8 in this near tetraploid subclone.

48,XX,t(12;16)(q13;p11.1),+17,+20[7]/96,sl×2,inv(3)(q21q27),t(3;6)(p25;q21)[19]
48,XX,t(12;16)(q13;p11.1),+17,+20[7]/96,idem×2,inv(3)(q21q27),t(3;6)(p25;q21)[19]
> The mainline with 96 chromosomes is a doubling product of the hyperdiploid stemline and has in addition an inv(3) and a balanced t(3;6), i.e., there are two normal chromosomes 3 and three normal chromosomes 6 in this near tetraploid subclone.

48,XX,t(12;16)(q13;p11.1),t(14;19)(q23;p11),+17,–19,+20,+21[32]/49,sl,+6[17]
48,XX,t(12;16)(q13;p11.1),t(14;19)(q23;p11),+17,–19,+20,+21[32]/49,idem,+6[17]
> The subclone with 49 chromosomes has all the abnormalities seen in the stemline plus an extra chromosome 6.

53,XY,+1,+12,+14,t(14;18)(q32;q21),+15,+16,+18,+20[21]/53,sl,del(7)(q21)[9]
53,XY,+1,+12,+14,t(14;18)(q32;q21),+15,+16,+18,+20[21]/53,idem,del(7)(q21)[9]
    The sideline has a deletion of the long arm of chromosome 7 in addition to the abnormalities seen in the stemline.

45,XY,t(1;6)(p34;q22),–3[13]/49,sl,+3,+del(7)(q11.2),+8,+9[22]
45,XY,t(1;6)(p34;q22),–3[13]/49,idem,+3,+del(7)(q11.2),+8,+9[22]
    There are four additional changes in the subclone with 49 chromosomes in relation to the stemline. Note, however, that the stemline has monosomy 3 whereas the sideline has two normal chromosomes 3, i.e., +3 in this situation does not denote that the clone has trisomy 3.

47,XX,inv(6)(p21q25),+12[17]/50,sl,–inv(6),+7,+8,+9,+mar[11]
47,XX,inv(6)(p21q25),+12[17]/50,idem,–inv(6),+7,+8,+9,+mar[11]
    The inv(6) present in the stemline has not been found in the sideline with 50 chromosomes. The breakpoints in the inv(6) need not be repeated. Note that there is monosomy 6 in the sideline. If the sideline was disomic for chromosome 6 then it would be written:
    50,sl,+6,–inv(6),+7,+8,+9,+mar[11] or 50,idem,+6,–inv(6),+7,+8,+9,+mar[11].

11.1.5 Composite Karyotype

In many instances, especially in solid tumors, there is great karyotypic heterogeneity within the tumor, but different cells nevertheless share some cytogenetic characteristics. *Every effort should be made to describe the subclones so that clonal evolution is made evident.* However, in some instances, a *composite karyotype* (**cp**) will have to be created. The composite karyotype contains all clonally occurring abnormalities and should also give the range of chromosome numbers in the metaphase cells containing the clonal abnormalities. The total number of cells in which the clonal changes were observed is given in square brackets after the karyotype, preceded by the symbol **cp**. The term **cp** should not be used to describe random loss; in an otherwise normal result, random losses or gains should be interpreted as consistent with technical artifact and should not be included in nomenclature. This is distinct from composite karyotype where clonal changes are identified in the context of nonclonal abnormalities.

47~55,XX,del(3)(p12),+i(6)(p10),del(7)(q11.2),+8,dup(11)(q13q25),+16,+17,der(18)t(18;20)(q23;q11.1),+21,+21,+22[cp24]
    Each of the abnormalities in this example has been seen in at least two cells, but there may be no cell with all abnormalities. The fact that it is a composite karyotype is obvious from the symbol cp and also because the chromosome number is given as a range.

It is not apparent from a composite karyotype how many cells have each abnormality. This information may be expressed by providing the number of cells in square brackets after each abnormality.

45~48,XX,del(3)(p12)[2],–5[4],+8[2],+11[3][cp7]
    In this composite karyotype, constructed on the basis of a total of seven cells, each with at least one of the abnormalities listed, two cells had a terminal deletion of the short arm of chromosome 3 with a breakpoint in 3p12, four cells had monosomy 5, two cells had trisomy 8, and three cells had trisomy 11. Some cells had more than one of these abnormalities.

It should be noted that in a composite karyotype the sum of the aberrations listed may indicate a higher or lower chromosome number than that actually seen. For example, if the following five cells are karyotyped

48,XX,+7,+9
48,XX,+7,+11
48,XX,+9,+11
48,XX,+9,+13
48,XX,+13,+21

then the composite karyotype will be

48,XX,+7,+9,+11,+13[cp5]
    The chromosome number in each of the five cells containing a clonal abnormality is 48, which is given as the chromosome number of the composite karyotype, although the sum total of all clonal changes indicates a chromosome number of 50. No cell with 50 chromosomes was observed.

Note also that a composite karyotype may contain such seemingly paradoxical abnormalities as loss and gain of the same chromosome. For example, if the following six cells are karyotyped

45,XX,−15,del(17)(q11.1)
46,XX,+7,−15,del(17)(q11.1)
46,XX,+12,−15
47,XX,+7
47,XX,+15,del(17)(q11.1)
48,XX,+12,+15

then the composite karyotype will be

45~48,XX,+7,+12,+15,−15,del(17)(q11.1)[cp6]
    Trisomy 15 and monosomy 15 are both clonal changes, present in two and three cells, respectively.

42,XX,−2,−16,−21,−22
44,XX,−1,−7,+8,−11
44,XX,−7,+8,−12,−13
43,XX,−7,−18,−20
46,XX,−7,+8

Composite karyotype:

43~46,XX,−7,+8[cp4]
    Note that the cell with 42 chromosomes is not included because the abnormalities seen are due to random loss and are not part of the clone.

51,XY,+1,−7,+8,t(9;22)(q34;q11.2),+11,+13,+19,+der(22)t(9;22)
51,XY,+1,+5,−7,+8,t(9;22)(q34;q11.2),+11,+19,+der(22)t(9;22)
51,XY,+1,+5,−7,+8,t(9;22)(q34;q11.2),+13,+19,+der(22)t(9;22)
52,XY,+1,+5,−7,+8,t(9;22)(q34;q11.2),+11,+13,+19,+der(22)t(9;22)
46,XY,t(9;22)(q34;q11.2)[5]

Composite karyotype:

46,XY,t(9;22)(q34;q11.2)[5]/51~52,sl,+1,+5,−7,+8,+11,+13,+19,+der(22)t(9;22)[cp4]
46,XY,t(9;22)(q34;q11.2)[5]/51~52,idem,+1,+5,−7,+8,+11,+13,+19,+der(22)t(9;22)[cp4]
> This is an example of a stemline with a 9;22 translocation. Multiple cells were found with gain and loss of chromosomes, likely due to clonal evolution. But because not all aberrations are present in all cells, these cells have been combined into a composite karyotype.

11.1.6 Unrelated Clones

Clones with completely unrelated karyotypic anomalies are presented according to their size; the largest first, then the second largest, etc. If there are two equal sized clones, they are listed as follows: clones with abnormalities of the sex chromosomes first and then those with the smallest to largest numbered autosomes. A normal diploid clone, when present, is always listed last.

46,XX,t(3;9)(p13;p13)[14]/48,XX,+3,+9[11]/46,XX,t(1;6)(p11;p12)[9]/47,XX,t(6;10)(q12;p15),+7[6]/46,XX,inv(6)(p22q23)[3]/46,XX[7]
> Five different clones in the same tumor presented in order of decreasing frequency, irrespective of chromosome number or type of aberration.

If a tumor contains both related and unrelated clones, the former are presented first in order of increasing complexity (see Section 11.1.4), followed by the unrelated clones in order of decreasing frequency.

50,XX,t(2;6)(p22;q16),+5,+5,+8,+11[19]/51,sl,+8[7]/52,sdl1,+9[12]/46,XX,del(3)(q13)[11]/47,XX,+7[6]/46,XX,t(1;3)(p22;p14)[4]
50,XX,t(2;6)(p22;q16),+5,+5,+8,+11[19]/51,idem,+8[7]/52,idem,+8,+9[12]/46,XX,del(3)(q13)[11]/47,XX,+7[6]/46,XX,t(1;3)(p22;p14)[4]
> The three related clones take precedence over the unrelated clones and are presented first. The three unrelated clones are presented in order of decreasing frequency.

However, if a previously identified abnormality is found among other unrelated clones, it should be listed first, regardless of the number of cells in which it is identified.

46,XY,t(9;22)(q34;q11.2)[6]/46,XY,t(1;3)(p22;p14)[14]

11.2 Modal Number

The *modal number* is the most common chromosome number in a tumor cell population. The modal number may be expressed as a range between two chromosome numbers.

Modal numbers in the *haploid* (**n**), *diploid* (**2n**), *triploid* (**3n**) or *tetraploid* (**4n**) range, or near but not equal to any multiple of the haploid number, and which cannot be given as a precise number, may be expressed as *near-haploid* (**n±**), *hypohaploid* (**n−**), *hyperhaploid* (**n+**), *near-diploid* (**2n±**), *hypodiploid* (**2n−**), *hyperdiploid* (**2n+**), *near-triploid* (**3n±**), *hypotriploid* (**3n−**), *hypertriploid* (**3n+**), *near-tetraploid* (**4n±**), *hypotetraploid* (**4n−**), *hypertetraploid* (**4n+**), and so on. Each range is determined as n±n/2, with n/2 defined operationally as 11 chromosomes. Suggested examples of ploidy levels, including ranges of chromosome numbers constituting each level, are given in the following table.

Near-haploidy (23±)	**≤34**	**Near-pentaploidy (115±)**	**104–126**
Hypohaploidy	<23	Hypopentaploidy	104–114
Hyperhaploidy	24–34	Hyperpentaploidy	116–126
Near-diploidy (46±)	**35–57**	**Near-hexaploidy (138±)**	**127–149**
Hypodiploidy	35–45	Hypohexaploidy	127–137
Hyperdiploidy	47–57	Hyperhexaploidy	139–149
Near-triploidy (69±)	**58–80**	**Near-heptaploidy (161±)**	**150–172**
Hypotriploidy	58–68	Hypoheptaploidy	150–160
Hypertriploidy	70–80	Hyperheptaploidy	162–172
Near-tetraploidy (92±)	**81–103**	**Near-octaploidy (184±)**	**173–195**
Hypotetraploidy	81–91	Hypooctaploidy	173–183
Hypertetraploidy	93–103	Hyperoctaploidy	185–195

Ploidy levels are recommended but exceptions may be made if biologically significant.

81<3n>,XXX,+X,+X,+X,+X,+X,+1,+1,+3,+3,+14,+14,+14,−15,+21
   A tumor in a female shows 81 chromosomes but is reported relative to triploid since most chromosomes are trisomic and this is biologically significant for the disease in question, although the count is in the near-tetraploid range.

Pseudodiploid, pseudotriploid, etc., are used to describe a karyotype, which has the number of chromosomes equal to a multiple of the haploid number (*euploid*) but is abnormal because of the presence of acquired numerical and/or structural aberrations. All chromosome numbers deviating from euploidy are *aneuploid*.

The description of sex chromosome abnormalities poses a special problem in male tumors with uneven ploidy levels (haploid, triploid, pentaploid, etc.) because the expected sex chromosome constitution cannot be deduced. For example, the sex chromosome constitution of a triploid tumor might theoretically be XXY or XYY. By convention, in males all sex chromosome deviations should be expressed in relation to X in haploid tumors, to XXY in triploid tumors, to XXXYY in pentaploid tumors, and so on:

68<3n>,XY,−X[10]
   A tumor in a male shows triploidy with loss of a sex chromosome.

11.3 Constitutional Karyotype

The same clonality criteria (see Section 11.1.1) apply to cells containing the constitutional karyotype as to cells containing acquired chromosome abnormalities. A normal diploid clone, when present, is always listed last.

A constitutional anomaly is indicated by the letter **c** after the abnormality designation. In the description of the karyotype, the constitutional anomaly is listed in chromosome number order (see Chapter 6). A clone with only a constitutional anomaly is, as the normal diploid clone, always listed last.

47,XXYc[5]/46,XY[15] or 47,XY,+X[5]/46,XY[15]
   Tumor cells in a male with either constitutional mosaicism for an extra X chromosome, or acquired gain of one X chromosome.

48,XX,+8,+21c[20]
: Tumor cells with a constitutional trisomy 21 and an acquired trisomy 8.

47,X,t(X;18)(p11.1;q11.1),+21c[20]
: Tumor cells with a constitutional trisomy 21 and an acquired t(X;18).

47,XXYc,t(9;22)(q34;q11.2)[20]
: Tumor cells with a constitutional XXY and an acquired t(9;22).

48,XY,+8,inv(10)(p12q22)mat,+21[20]
: Tumor cells with a constitutional inv(10) and acquired trisomies 8 and 21. Note that when the inheritance is known, the mat or pat takes the place of the c.

47,XX,del(5)(q15),+mar c[20]
: Tumor cells with a constitutional marker chromosome of unknown origin and an acquired deletion of the long arm of one chromosome 5. For constitutional markers, there is a space between **mar** and **c**.

48,XY,+8,+21c[3]/49,idem,+9[5]/47,XY,+21c[12]
: Tumor cells with a constitutional trisomy 21 and acquired trisomies 8 and 9. The clone with only the constitutional trisomy 21 is listed last irrespective of the size of this clone.

49,XX,t(2;13)(q37;q14),+18c,+18,+mar[3]/47,XX,+18c[17]/46,XX[8]
: Tumor cells with a constitutional mosaic trisomy 18 as the sole anomaly in one clone and with acquired abnormalities, including an additional chromosome 18, in another clone. The clone with 49 chromosomes thus has four chromosomes 18. The clone with only the constitutional trisomy 18 is listed next. The normal cell line is listed last.

To describe acquired abnormalities affecting one of the chromosomes of a pair that is involved in a constitutional anomaly, the constitutional aberration must always be given, even if none of the tumor cells have this particular aberration. Thus, an acquired abnormality is always presented in relation to the constitutional karyotype.

46,XX,+21c,−21[20]
: The patient has a constitutional trisomy 21 and the acquired abnormality in the tumor cells is a loss of one chromosome 21.

45,Xc,t(X;18)(p11.1;q11.1)[20]
: Tumor cells in a patient with Turner syndrome (45,X) have an acquired t(X;18), i.e., the only X chromosome is involved in the translocation and consequently there is no normal X chromosome in the tumor cells.

46,XX,der(9)t(9;11)(p22;q23)t(11;12)(p13;q22)c,der(11)t(9;11)t(11;12)c,der(12)t(11;12)c[20]
: Female patient with a known constitutional t(11;12)(p13;q22) presents with t(9;11) positive AML. The derivative chromosome 11 involved in the t(11;12)c is also involved in the t(9;11) aberration. The resulting karyotype, with both constitutional and acquired aberrations, should list each aberrant chromosome as a derivative chromosome.

11.4 Counting Chromosome Aberrations

The recommended method of counting of chromosome abnormalities is outlined below. Only clonal abnormalities are counted. When multiple clones are present, each independent aberration is counted only once. There is no attempt to define complexity as there are disease-specific definitions which are used for cytogenomic prognostic risk assessment (Chun et al., 2010; Grimwade et al., 2010; Baliakas et al., 2019; Haase et al., 2019). To determine complexity in a specific disease entity, it is mandatory to count chromosome aberrations in the same way as in the corresponding study on which the prognostic system is based.

Abnormality type	Examples	Number of chromosome abnormalities
Numerical gain	Trisomy Duplication of a derivative chromosome	1
Numerical loss	Monosomy, includes –Y	1
Balanced structural abnormality *(no gain, no loss of chromosomal material)*	Simple balanced translocation Complex balanced translocation (involving 3 or more chromosomes) Inversion Balanced insertion	1
Unbalanced aberrations involving one chromosome (leading to gain or loss of chromosomal material)	Isochromosome Deletion* Duplication* Simple ring chromosome Isodicentric chromosome Homogeneously staining region Double minutes Unidentified marker chromosome	1
	Tetrasomy of same chromosome Triplication or quadruplication Isoderivative chromosome	2
Unbalanced aberrations involving two or more chromosomes	Unbalanced translocation Unbalanced insertion Derivative chromosome Complex ring chromosome Isoderivative chromosome	2
Ploidy abnormalities	Multiplication of complete chromosome set (normal or aberrant)	1
Multiple clones (subclones or independent clones)	Count chromosome abnormalities in each clone/subclone separately Number of chromosome abnormalities is defined by the clone with highest number of abnormalities *For composite karyotype* Count clonal chromosomal aberrations in metaphase with highest number of abnormalities	
Constitutional abnormalities	Do not include in the count; if unknown then include	0

* Includes multiple of one chromosome.

12 Meiotic Chromosomes

During late prophase-first metaphase, the bivalents may be grouped by size, and bivalent 9 can sometimes be distinguished by its secondary constriction. At these stages, the Q- and C-staining methods are particularly informative. The autosomal bivalents generally show the same Q-band patterns as somatic chromosomes. The C-staining method reveals the centromere position, thus allowing identification of the bivalents in accordance with the conventionally stained somatic chromosomes. There are, however, minor differences in the C-band patterns between the bivalents and mitotic chromosomes.

When the Q- and C-staining methods are used consecutively, further distinction of the bivalents is possible. Measurements of the relative length of orcein-stained bivalents, previously identified by these special techniques, are in good agreement with corresponding mitotic measurements. Chiasma frequencies have been determined for individual bivalents.

The Y chromosome can be identified at all meiotic stages by the intense fluorescence of its long arm. Both the Q- and C-staining methods have revealed that the short arm of the Y is associated with the short arm of the X in the first meiotic metaphase.

12.1 Terminology

The symbols **PI**, **MI**, **AI**, **MII**, and **AII** are used to indicate the stage of meiosis, namely, *prophase of the first division, first metaphase* (including diakinesis), *first anaphase, second metaphase*, and *second anaphase*. This is followed by the total count of separate chromosomal elements. The sex chromosomes are then indicated by XY or XX when associated and as X,Y when separate. Any additional, missing, or abnormal element follows, with that element specified within parentheses and preceded by the Roman numeral **I**, **II**, **III**, or **IV** to indicate if it is a *univalent, bivalent, trivalent,* or *quadrivalent*, respectively. The absence of a particular element is indicated by a **minus** (–) sign. The **plus** (+) sign is used in first metaphase only when the additional chromosome is not included in a multivalent. The chromosomes involved in a rearrangement are listed numerically within parentheses and separated by a **semicolon** (;).

A more detailed description, for instance, of the chromosomal segments involved in a rearrangement may be included within parentheses using the standard nomenclature, with which this meiotic notation has been designed to conform. When necessary, use of the symbols **fem** and **mal** is recommended for *female* and *male*, respectively, and when a more detailed description of different premeiotic and meiotic stages is required, the following symbols may be used:

spm Spermatogonial metaphase
oom Oogonial metaphase
lep Leptotene
zyg Zygotene
pac Pachytene
dip Diplotene
dit Dictyotene
dia Diakinesis

The symbol **xma** is suggested for *chiasma(ta)*. The total number of chiasmata in a cell can be designated by placing this symbol, followed by an **equal sign** (=) and a two-digit number, in parentheses, e.g., (xma=52). In the case of a meiotic cell with a low number of chiasmata, a single digit should be preceded by a zero, e.g., (xma=09).

The number of chiasmata in a bivalent or multivalent or their arms may be indicated by a single digit, e.g., (xma=4).

Location of chiasmata can be indicated by the standard arm symbols **p** and **q**, supplemented by **prx** for *proximal*, **med** for *medial*, **dis** for *distal*, and **ter** for *terminal*. The band or region number can be used when such precise information is available.

Chromosomes participating in a bivalent or multivalent are specified within parentheses after the Roman numeral that describes the bivalent (II) or the type of multivalent (III, IV, etc.). If the number of chiasmata within the multivalent is known, this is indicated within parentheses in consecutive order, i.e., the number of chiasmata between the first and second chromosome is given first, between the second and the third next, etc. The last figure then indicates the number of chiasmata between the last and first chromosome. If the number of chiasmata in non-interstitial and interstitial segments can be specified separately, these should be represented by a **plus** (+) sign. The number of chiasmata in the non-interstitial segment is written first, e.g., (xma=2+1), indicating two chiasmata in the non-interstitial and one in the interstitial segment. It is assumed that a careful description of the mitotic karyotype of the subject will be given separately.

12.1.1 Examples of Meiotic Nomenclature

MI,23,XY
   A primary spermatocyte at diakinesis or metaphase I with 23 elements, including an XY bivalent.

MI,24,X,Y
   A primary spermatocyte at diakinesis or metaphase I with 24 elements, including X and Y univalents.

MI,23,XY,III(21)
   A primary spermatocyte with 23 elements from a male with trisomy 21. The three chromosomes 21 are represented by a trivalent.

MI,24,XY,+I(21)
   A primary spermatocyte with 24 elements from a male with trisomy 21. The extra chromosome 21 is represented by a univalent.

MI,22,XY,III(13q14q)
A primary spermatocyte with 22 elements from a der(13;14)(q10;q10) heterozygote. The translocation chromosome is represented by a trivalent.

MI,23,XY,(xma=52)
Spermatocyte in first metaphase with 23 elements, including an XY bivalent. The total number of chiasmata in the cell is 52, the association between the X and Y chromosomes being counted as one chiasma.

fem dia,II(2,2)(xma=4)
Oocyte in diakinesis in which bivalent 2 has four chiasmata.

fem dia,II(2,2)(xma=4)(p=2,q=2)
Female diakinesis in which bivalent 2 has four chiasmata. The positions of the chiasmata are known. Thus (xma=4)(p=2,q=2) indicates that there are two chiasmata on the short arm and two on the long arm. More precise location of the chiasmata could then be indicated, e.g., by (xma=4) (pter,pprx,qmed,qdis). Alternatively, if the chiasmata have been localized to specific regions, these could be indicated, e.g., by (xma=4)(pter,p1,q2,qter).

mal MI,III(14,14q21q,21)(xma=3)
Male first metaphase with a trivalent composed of one chromosome 14, one 14q21q Robertsonian translocation chromosome, and one chromosome 21. There are three chiasmata, the positions of which have not been specified.

MI,23,X,Y,III(13,13q14q,14)(xma=2,1),(xma=51)
Spermatocyte in first metaphase with 23 elements, univalent X and Y chromosomes, and one trivalent composed of one chromosome 13, one 13q14q Robertsonian translocation chromosome, and one chromosome 14. There are two chiasmata between the normal chromosome 13 and the 13q14q translocation chromosome and one chiasma between the translocation chromosome and the normal chromosome 14. Altogether, there are 51 chiasmata in the cell.

fem dia,IV(2,der(2),5,der(5))(xma=2+1,1,1+0,1)
Oocyte in diakinesis with a quadrivalent composed of two normal chromosomes and two derivative chromosomes of chromosomes 2 and 5, respectively. There are three chiasmata between chromosomes 2 and der(2), of which two are in the non-interstitial segment and one is in the interstitial segment. In addition, there is one chiasma between der(2) and chromosome 5, one in the non-interstitial and none in the interstitial segment between chromosome 5 and der(5), and finally one between der(5) and chromosome 2. The last chiasma indicates that the quadrivalent has a ring shape.

MI,24,X,Y,III(2,der(2),5)(xma=4),I(der(5)),(xma=51)
Spermatocyte in first metaphase with 24 elements, including univalent X and Y chromosomes, one trivalent, and one additional univalent. The trivalent is composed of one normal and one derivative chromosome 2, as well as one normal chromosome 5. This trivalent has a total of four chiasmata, the positions of which are not known. One univalent is composed of one derivative chromosome 5. The total number of chiasmata in the cell is 51.

MII,22,X,−16,+16cht,+16cht
Oocyte at second meiotic metaphase in which chromosome 16 is absent, but is represented by its two single chromatids.

12.1.2 Correlation between Meiotic Chromosomes and Mitotic Banding Patterns

Meiotic chromosomes from pachytene spermatocytes have been shown to exhibit chromomere patterns without any pretreatment, which correspond well with Giemsa-dark bands of somatic chromosomes, suggesting that both represent a basic structural feature of the mammalian chromosome. Just as in the case of somatic chromosome bands, the number of chromomeres that can be recognized is a function of the stage of contraction. The chromomere patterns of human oocyte pachytene chromosomes are apparently similar to those of spermatocyte chromosomes, although the former exhibit less contraction and hence more chromomeres.

Figure 10 is an idiogram of the 22 autosomal bivalents from pachytene spermatocytes employing the nomenclature of the somatic chromosome banding patterns using the 850-band stage nomenclature because this stage of contraction corresponds approximately to mid-pachytene of spermatocyte meiosis. The chromomeres are given the numbers of Giemsa-positive bands and inter-chromomere regions are given the numbers of Giemsa-negative bands. A special feature of human pachytene chromosomes is the presence of a particulate or puff-like structure located at the heterochromatic region (9q12). The structure is transient and is limited to the pachytene stage.

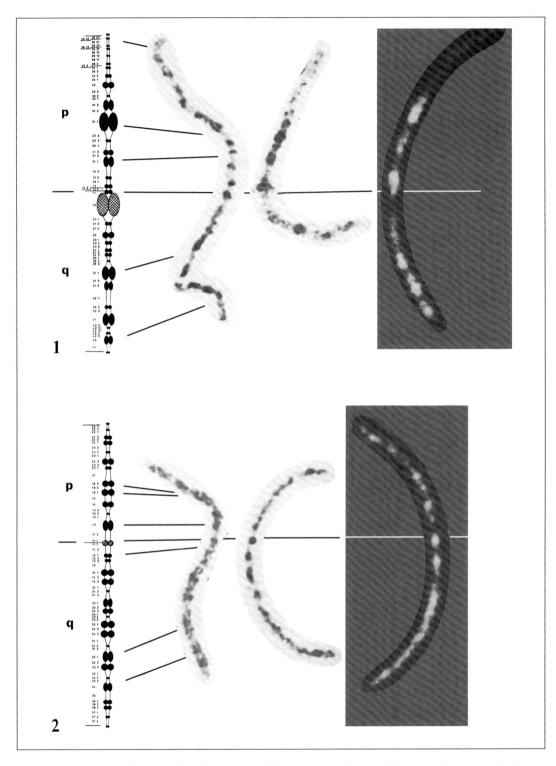

Fig. 10. Chromomere idiogram of the 22 autosomal bivalents at pachytene. The nomenclature used is that of the 850-band stage of somatic chromosomes (ISCN 1981). Chromomere numbers are equivalent to those of Giemsa-positive bands of somatic chromosomes and inter-chromomere region numbers are equivalent to those of Giemsa-negative bands of somatic chromosomes. Individual bivalent idiograms compared with photomicrographs of bivalents (two Giemsa-stained and one quinacrine-stained). Lines between idiograms and photomicrographs connect centromeres and chromomeres that correspond to landmark bands of somatic chromosomes. (Method of Jhanwar et al., 1982; courtesy of Drs. R.S.K. Chaganti and S.C. Jhanwar).

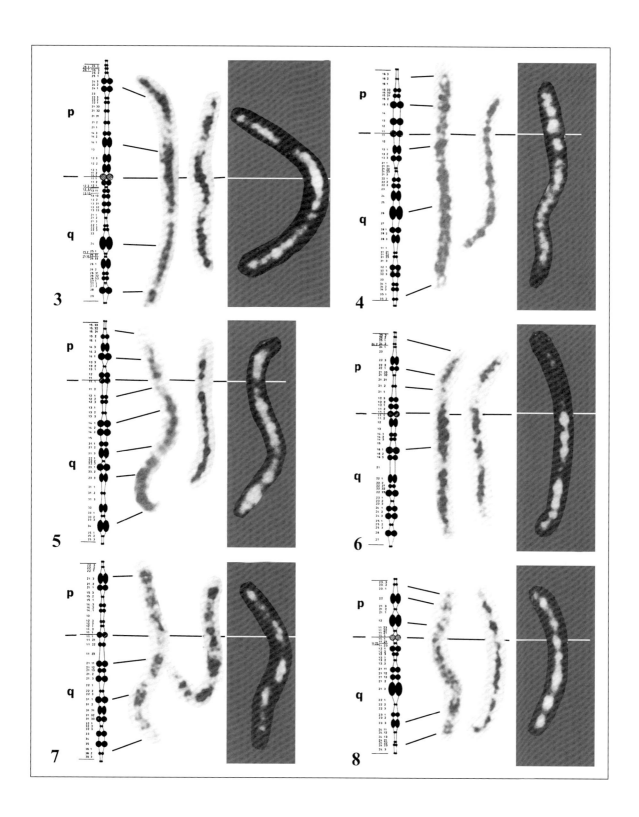

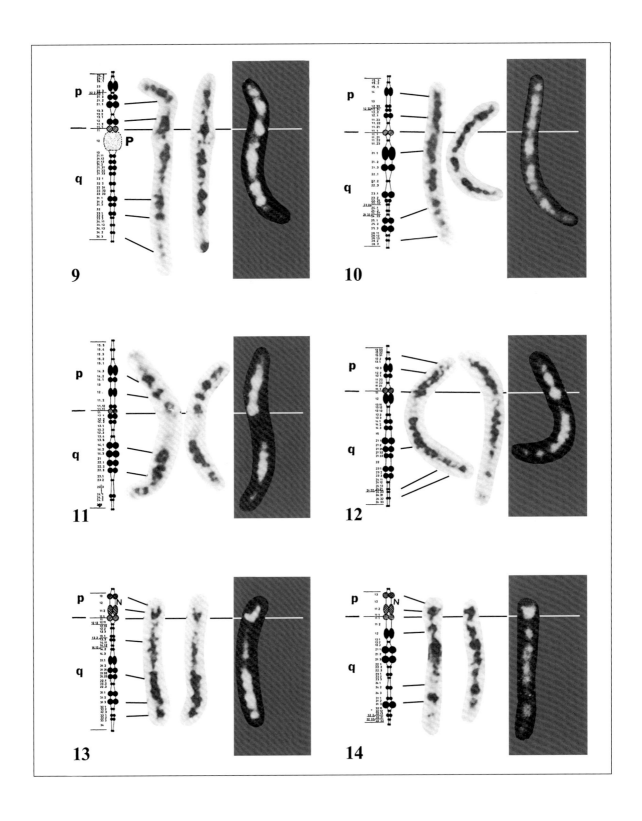

Fig. 10. continued (see legend on p 442)

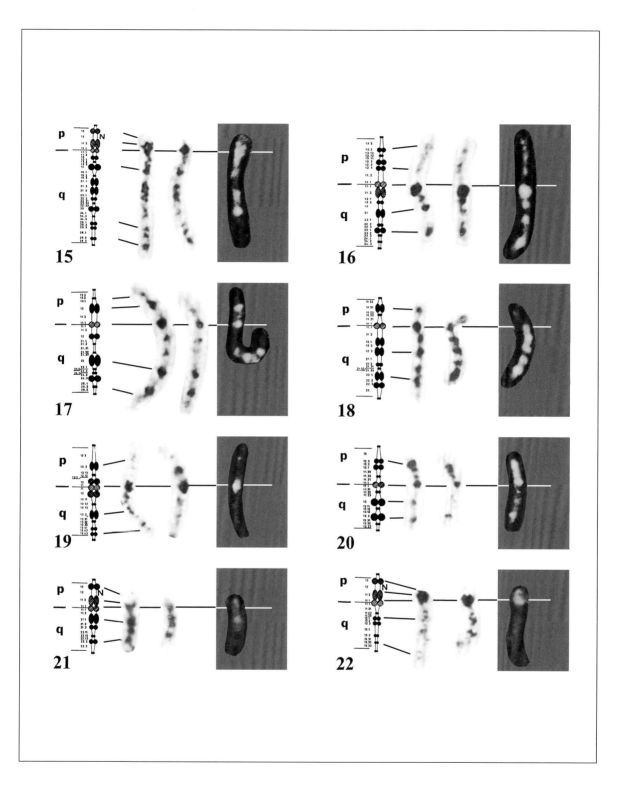

13 *In situ* Hybridization

13.1 Introduction

Major advances in human cytogenomics since the publication of ISCN (1985) include the development and implementation of a variety of non-isotopic *in situ* hybridization techniques to detect (Lichter et al., 1990; Trask, 1991), and in some instances quantify (Kallioniemi et al., 1992), specific DNA sequences and to locate them to specific chromosomal sites. The ever-increasing availability of a number of sequence-specific DNA probes, their amplification by the polymerase chain reaction, and the availability of fluorochrome-tagged reporter molecules that are bound to DNA probes, have all contributed to bridging the gap between the microscope and the molecule.

Techniques utilizing fluorescence *in situ* hybridization (FISH) allow the use of a number of fluorochromes so that the locations of different and differently tagged probes, and the relative positions of their binding sites, may be visualized microscopically on a single chromosome segment or DNA/chromatin fiber (Wiegant et al., 1993). In addition, the use of composite probes, coupled with suppressive hybridization (Landegent et al., 1987), enables whole chromosomes, or chromosome segments, to be specifically "painted" and uniquely visualized (Lichter et al., 1988; Pinkel et al., 1988; Guan et al., 1994). FISH banding methods are available and reviewed in Liehr et al. (2006). These developments have also enabled the cytogeneticist to detect the presence of specific DNA sequences in interphase nuclei and to visualize their distribution (Cremer et al., 1986). FISH applied to free, linearly extended chromatin fibers or naked DNA strands has increased the resolution of FISH interphase mapping to ≤1 kb (Wiegant et al., 1992; Parra and Windle, 1993).

FISH techniques have provided the cytogeneticist with an increased ability to identify chromosome segments, to correlate chromosome structures with gene locations, to reveal cryptic abnormalities that are undetectable using standard banding techniques, and to analyze and describe complex rearrangements.

By convention, fusion genes decribed in text are indicated by use of a hyphen (e.g., *BCR-ABL1*) while FISH probe cocktails are described using slashes (e.g., BCR/ABL1).

Where multiple techniques are used, the karyotype is listed first, followed by FISH, followed by results obtained with other techniques, each separated by a period (.). Alternatively, results obtained using different techniques may be presented on separate lines without periods.

13.2 Prophase/Metaphase *in situ* Hybridization (ish)

If a standard cytogenetic observation has been made, it may be given followed by a **period** (.), the symbol **ish**, a space, and the ish results. If a standard cytogenetic observation has not been made, the ish observations only are given.
- If FISH further clarifies the karyotype and, in retrospect, the abnormality can be visualized with banding, the karyotype may be re-written to reflect this new FISH information. If the abnormality is cryptic and cannot be visualized by banding, the abnormality should not be listed in the banded karyotype.
- The locus designations (in capital letters but not in italics) are separated by commas and the status of each locus is given immediately after the locus designation.
- When available, the clone name is preferred.
- If the clone name is not available, the locus designated according to either the UCSC or Ensembl Genome Browsers (www.genome.ucsc.edu/ and www.ensembl.org/) should be used in order as they would appear on the chromosome being described from pter to qter.
- If no locus is available, the gene name can be used, according to HUGO-approved nomenclature (www.hugo-international.org/).
- Although gene acronyms are usually italicized, they should not be italicized in the nomenclature.
- At the discretion of the investigator or laboratory director the probe name, clone name, accession number, gene name, or D-number can be used.
- When contig probes are used each locus may be listed, separated by **single slant lines (/)** or a single designation may be used in the nomenclature and the composition is described in the report.
- The band designation for the locus should be based on the current UCSC Genome Browser.
- Observations on structurally **abnormal** chromosomes are expressed by the symbol **ish**, followed by a space and then the symbol for the structural abnormality (whether seen by standard techniques and ish or only by ish), followed in separate parentheses by the chromosome(s), the breakpoint(s), and the locus or loci for which probes were used. **Presence (+)** or **absence (−)** is indicated within the same parentheses as the locus designation. When the number of signals on an abnormal chromosome can be counted, this may be indicated by multiple "+" symbols. When FISH results are used to clarify or extend the breakpoints identified in banded chromosomes, these are presented in the ish description.
- Observations on **normal** chromosomes are expressed by the symbol **ish** followed by a space and the chromosome, region, band, or sub-band designation of the locus or loci tested (not in parentheses), followed in parentheses by the locus (loci) tested, a **multiplication sign (×)** and the number of signals seen.
- The breakpoints need not be given in the FISH nomenclature unless it clarifies or extends the breakpoints given in the G-banded nomenclature.

13.2.1 Normal Signal Pattern

46,XX.ish X(DXZ1×2,SRY×0)
   A normal female. FISH with probes for the X centromere and the *SRY* gene was performed and no evidence of *SRY* was found. Where the clinical question is whether *SRY* is present or not, it is appropriate to indicate the status in the nomenclature.

46,XY.ish 4p16.3(D4F26,D4S96)×2
   A normal male [e.g., father of a child with der(4)t(4;11)(p16.3;p15)dmat] was tested by ish using probes for loci D4F26 and D4S96. There were two appropriately located copies of both probes.

46,XY.ish 17p11.2(RAI1×2)
: The *RAI1* locus on chromosome 17 is present in the normal copy number (two copies) as determined by ish with a locus-specific probe.

46,XX.ish 17p11.2(RAI1×2),21q21.3(D21S259/D21S341/D21S342×2)
: A normal female. The *RAI1* locus on chromosome 17 and the Down syndrome critical region on chromosome 21 are both present in the normal copy number (two copies) as determined by ish with a locus-specific probe. All loci of the contig probe are listed.

46,XY.ish 22q11.2(D22S75×2)
: Conventional cytogenetic analysis showed a normal male karyotype, and FISH using a probe in the DiGeorge syndrome region (D22S75) showed a normal hybridization pattern on metaphase chromosomes. The control probe is not given.

13.2.2 Abnormal Signal Patterns with Single Probes

In these FISH examples either the clinically relevant probe or the informative control probe has an abnormal signal pattern. The control probe is not given if it has a normal signal pattern.

46,XX.ish der(X)t(X;Y)(p22.3;p11.3)(SRY+)
: A presumed unbalanced translocation between the X and Y short arms resulting in a derivative X containing *SRY* at Xp22.3.

46,XX,ins(2)(p13q21q31).ish ins(2)(wcp2+)
: A direct insertion of the long-arm segment 2q21q31 into the short arm at band 2p13 was confirmed as derived from chromosome 2 by ish using whole chromosome paint 2.

46,XY,ins(5;2)(p14;q32q22).ish ins(5;2)(wcp2+)
: An inverted insertion of a chromosome 2 segment into the short arm of chromosome 5 was confirmed as derived from chromosome 2 using whole chromosome paint 2.

46,XX,t(2;17)(q32;q24)[20].ish t(2;17)(AC005181+;AC005181+)[20]
: Translocation disrupts the region corresponding to BAC clone AC005181, at 17q24, resulting in a signal on both derivative chromosomes. Cell numbers are given as this is an oncology sample.

46,XX.ish del(7)(q11.23q11.23)(ELN–)
: A female with a normal karyotype by cytogenetic analysis has a microdeletion in the Williams syndrome region of chromosome 7 identified by ish with an elastin gene (*ELN*) probe.

47,XY,+mar.ish der(8)(D8Z1+)
: An extra marker chromosome identified by ish to be derived from chromosome 8 using an 8-specific alpha-satellite probe.

46,XY.ish dup(17)(p11.2p11.2)(RAI1++)
: The region containing the *RAI1* locus on chromosome 17 is duplicated as detected by ish on metaphase chromosomes. There is one signal in the homologous chromosome 17, not indicated in the karyotype.

46,XX[20].ish inv(21)(q11.2q22.1)(q11.2)(RUNX1+)(q22.1)(RUNX1−)[5]
  A cryptic inversion of the segment 21q11.2 to 21q22.1 was identified by ish using a probe for the *RUNX1* locus. Note that the inversion breakpoints are in separate parentheses to make the FISH information apparent.

46,XX.ish del(22)(q11.2q11.2)(D22S75−)
  A female with a normal karyotype by cytogenetic analysis has a deletion in the DiGeorge syndrome critical region (DGCR) on chromosome 22 identified by ish using a probe for locus D22S75.

ish del(22)(q11.2q11.2)(D22S75−)
  Conventional cytogenetic analysis was not performed but a deletion in the DGCR on chromosome 22 was identified by ish using a probe for locus D22S75.

ish del(22)(q11.2q11.2)(D22S75−),del(22)(q11.2q11.2)(D22S75−)
  Conventional cytogenetic analysis was not performed but a deletion in the DGCR in both chromosomes 22 was identified by ish using a probe for locus D22S75.

ish del(22)(q13.3q13.3)(ARSA−)
  Conventional cytogenetic analysis was not performed but an interstitial deletion of distal 22q was identified by ish using a probe to the *ARSA* locus.

13.2.3 Abnormal Signal Patterns with Multiple Probes

46,X,r(X).ish r(X)(p22.3q21)(KAL+,DXZ1+,XIST+,DXZ4−)
  A ring X was further defined by ish as containing the short arm marker *KAL1*, the X alpha-satellite DXZ1 and the *XIST* gene on the long arm. It does not include DXZ4 at Xq24.

46,X,+r.ish r(X)(p21q21)(wcpX+,DXZ1+)
  A ring chromosome was identified by ish as a derivative X chromosome using whole chromosome paint X and X alpha-satellite probe DXZ1.

46,XX.ish X(DXZ1×2),der(7)(q3?)(SRY+)
  FISH with probes for the X centromere and the *SRY* gene was performed, and the *SRY* gene was found to be located in one of the chromosomes 7, in band q3, sub-band not known.

46,X,?i(Y)(p10).ish idic(Y)(q11)(DYZ3+,DYZ1−,DYZ3+)
  A presumed isochromosome for the short arm of Y was shown by ish to have two centromeres and no heterochromatin.

46,XX,t(2;17)(q32;q24)[20].ish t(2;17)(CTD-3115L8+;RP11-959M22+,CTD-3115L8−)[20]
  A translocation between the long arms of chromosomes 2 and 17. The breakpoint has been located between BAC clone RP11-959M22 and CTD-3115L8 on chromosome 17, which results in CTD-3115L8 moving to chromosome 2 and RP11-959M22 being retained on chromosome 17. Cell numbers are given for an oncology sample.

46,XX,add(4)(p16.3).ish dup(4)(p16.3p15.2)(wcp4+,RP11-1076P8+,WHS+,
RP11-1076P8+,WHS+)
  A chromosome 4 has extra material attached at band 4p16.3. Utilizing a whole chromosome paint 4, the extra chromatin was identified as a duplicated region of chromosome 4, determined by a combination of FISH and G-banding to be 4p16.3 to 4p15.2. The duplicated chromosome contains two copies of the region covered by the RP11-1076P8 BAC probe and a probe for Wolf-Hirschhorn syndrome in the original orientation relative to 4pter. Note that the order of probes is given according to their relative position on the derivative chromosome. The breakpoints are given in the FISH nomenclature as they are further clarified relative to those obtained by banding.

46,XX,add(4)(p16.3).ish dup(4)(p15.2p16.3)(wcp4+,RP11-1076P8+,WHS+,WHS+,
RP11-1076P8+)

> A chromosome 4 has extra material attached at band 4p16.3. Utilizing a whole chromosome paint 4, the extra chromatin was identified as a duplicated region of chromosome 4, determined by FISH to be 4p16.3 to 4p15.2. The duplicated chromosome contains two copies of the region covered by the RP11-1076P8 BAC probe and a probe for Wolf-Hirschhorn syndrome. The duplicated region is in the reverse orientation relative to 4pter.

46,XX,add(4)(q31).ish der(4)dup(4)(q31q34)(wcp4+)add(4)(q34)(wcp4–)

> A chromosome 4 has extra chromatin attached at band 4q31. Using whole chromosome paint 4 the proximal part of the additional material was shown to be derived from chromosome 4. G-banding suggested a duplication of bands 4q31 to 4q34. However, there was additional material distal to the duplication which did not hybridize with whole chromosome 4 paint, and is therefore of unknown origin.

46,XX.ish t(4;11)(p16.3;p15)(wcp11+,D4F26–,D4S96+,D4Z1+;D4F26+,wcp11+)

> A cryptic reciprocal translocation between chromosomes 4 and 11 was identified by ish. The der(4) was positive with whole chromosome paint 11, a probe for D4S96 (Wolf-Hirschhorn region) and 4 alpha-satellite but negative for D4F26 (4p telomere region). The der(11) was positive for D4F26 as well as whole chromosome paint 11.

46,XX.ish der(4)t(4;11)(p16.3;p15)(wcp11+,D4F26–,D4S96+,D4Z1+)dmat

> This child is an unbalanced offspring from the segregation of the cryptic translocation above. She has one normal chromosome 4 and two normal chromosomes 11. The ish results of the der(4) are the same as the der(4) of the mother.

47,XY,+der(4)t(4;11)(p34;q13),t(4;11)[20].ish der(4)(3′KMT2A+),t(4;11)(3′KMT2A+;
5′KMT2A+,3′KMT2A–)[5]

> A male karyotype with a t(4;11) plus an extra copy of the der(4). There is a 3′KMT2A signal on both copies of the der(4). There is a 5′KMT2A signal on the der(11) component of the t(4;11).

46,XY,t(9;22)(q34;q11.2)[20].ish t(9;22)(ABL1–;BCR+,ABL1+)[5]

> A male karyotype with a t(9;22) that has been characterized by ish using a single-fusion probe. The probe sequence from the *ABL1* locus is missing from the derivative chromosome 9 and is present on the derivative chromosome 22 distal to the *BCR* locus.

47,XY,t(9;22)(q34;q11.2),+der(22)t(9;22)[20].ish t(9;22)(ABL1–;BCR+,ABL1+),der(22)
(BCR+,ABL1+)[5]

> A male karyotype with a t(9;22) plus an extra copy of the der(22) that has been characterized by ish using a single-fusion probe. The *ABL1* locus is missing from the derivative chromosome 9 and is present on both derivative chromosomes 22 distal to the *BCR* locus.

46,XX,t(9;22)(q34;q11.2)[10].ish t(9;22)(ABL1+,BCR+;BCR+,ABL1+)[5]

> A female karyotype with a t(9;22) detected using dual-fusion probes for *BCR* and *ABL1*. One copy of *ABL1* and one copy of *BCR* are found on each derivative chromosome.

46,XX,t(9;22)(q34;q11.2)[20].ish der(9)t(9;22)del(9)(q34q34)(ABL1–,BCR+),der(22)
t(9;22)(BCR+,ABL1+)[5]

> A female karyotype with a t(9;22) detected using dual-fusion probe for *BCR* and *ABL1*. There is a deletion on the derivative 9, encompassing the *ABL1* locus, not detected using conventional cytogenetic analysis.

46,XX,t(9;22)(q34;q11.2)[20].ish der(9)t(9;22)del(9)(q34q34)(ASS1–,ABL1–,BCR+),
der(22)t(9;22)(BCR+,ABL1+)[5]
 A female karyotype with a t(9;22) detected using a three-color dual-fusion probe for ASS1, ABL1, and BCR. There is a deletion on the derivative 9, encompassing the ASS1 and ABL1 loci, not detected using conventional cytogenetic analysis.

46,XX,t(9;22;10)(q34;q11.2;q22)[21].ish der(9)t(9;22;10)del(9)(q34q34)(wcp9+,
ABL1–,wcp10+),der(10)t(9;22;10)del(22)(q11.2q11.2)(wcp10+,BCR–,wcp22+),der(22)
t(9;22;10)(wcp22+,BCR+,ABL1+)[10]
 A female karyotype with a three-way t(9;22;10) detected using dual-fusion probe for BCR and ABL1. There is a single BCR-ABL1 fusion signal and loss of an ABL1 and a BCR signal. Whole chromosome paint analysis with probes specific for chromosome 9, 10, and 22 confirmed the presence of the t(9;22;10).

47,XX,t(9;22;10)(q34;q11.2;q22),+der(22)t(9;22;10)[20].ish der(9)t(9;22;10)del(9)(q34q34)
(ABL1–),der(10)t(9;22;10)del(22)(q11.2q11.2)(BCR–),der(22)t(9;22;10)(BCR+,ABL1+)[5]
 A female karyotype with a three-way t(9;22;10) and an additional copy of the derivative chromosome 22 detected using dual-fusion probe for BCR and ABL1. There is a BCR-ABL1 fusion signal on both copies of the derivative chromosome 22, with loss of an ABL1 signal from the derivative chromosome 9 and a BCR signal from the derivative chromosome 10.

46,XX[20].ish t(12;21)(p13;q22)(ETV6+,RUNX1+;RUNX1+,ETV6+)[5]
 Normal female chromosomes in an oncology sample which was also tested by metaphase FISH using a dual-fusion probe and found to have an ETV6-RUNX1 gene fusion from a cryptic reciprocal translocation between chromosomes 12 and 21.

45,XY,der(14;21)(q10;q10).ish dic(14;21)(p11.2;p11.2)(D14Z1/D22Z1+;D13Z1/D21Z1+)
 A Robertsonian translocation, der(14;21), is shown to be dicentric using ish. The breakpoints are given in the FISH nomenclature as they are further clarified relative to those obtained by banding.

46,XX,del(15)(q11.2q13).ish del(15)(SNRPN–,D15S10–)
 A cytogenetically detected deletion of bands 15q11.2q13 characterized by ish. Two loci (SNRPN and D15S10) from the Prader-Willi/Angelman region are deleted.

46,XY.ish del(15)(q11.2q11.2)(SNRPN–,D15S10–)
 A male with a normal karyotype by cytogenetic analysis has a microdeletion of the Prader-Willi/Angelman region of chromosome 15 identified by ish. The deletion includes the region defined by probes for the SNRPN and D15S10 loci.

46,XY.ish del(15)(q11.2q12)(D15S11+,SNRPN–,D15S10–,GABRB3+)
 A microdeletion of chromosome 15 defined by ish using probes for loci D15S11, SNRPN, D15S10 and GABRB3. SNRPN and D15S10 are deleted while D15S11 and GABRB3 are retained.

47,XY,+dic(15;15)(q11.1;q11.1).ish dic(15;15)(D15Z1+,D15Z4+,SNRPN–;SNRPN–,
D15Z4+,D15Z1+)
 A supernumerary dicentric chromosome 15 was shown by metaphase FISH to be positive for D15Z1 and D15Z4, but negative for SNRPN.

46,XX[20].ish ins(15;17)(q22;q21q21)(PML+,RARA+;RARA+)[5]
 A cryptic insertion of the segment 17q21 from the long arm of chromosome 17 into the 15q22 band of the long arm of chromosome 15 identified using probes for PML and RARA.

46,XX,inv(16)(p13.1q22)[20].ish inv(16)(p13.1)(5′CBFB+)(q22)(3′CBFB+)[5]
Inversion of chromosome 16 separates the two probes for the *CBFB* locus into the 5′ probe on the short arm and the 3′ probe on the long arm. Note in this and the following examples that the inversion breakpoints are in separate parentheses to make the FISH information apparent.

46,XX,inv(16)(p13.1q22)[20].ish inv(16)(p13.1)(RP11-620P11+)(q22)(RP11-620P11+)[5]
Inversion of chromosome 16 separates the region corresponding to BAC probe RP11-620P11 giving a signal on the short arm and on the long arm.

46,XX,inv(16)(p13.1q22)[20].ish inv(16)(p13.1)(MYH11+,CBFB+)(q22)(MYH11+,CBFB+)[5]
Pericentric inversion in which breakage and reunion have occurred at bands 16p13.1 and 16q22 is confirmed by probes for *MYH11* and *CBFB*.

46,XX,t(16;16)(p13.1;q22)[20].ish t(16;16)(3′CBFB+;3′CBFB–)[5]
Translocation disrupts the *CBFB* locus resulting in translocation of the 3′ probe from 16q22 to 16p13.1 on the other homologue.

47,XX,+mar.ish add(16)(p or q)(wcp16+,D16Z1+)
An extra marker chromosome identified by ish to be partially derived from chromosome 16 (arm unknown) using whole chromosome paint 16 and the 16 alpha-satellite probe. Additional material of unknown origin is present in the marker.

47,XY,+mar.ish der(17)(wcp17+,D17Z1+)
An extra marker chromosome identified by ish as derived from chromosome 17 using whole chromosome paint 17 and a 17-specific alpha-satellite probe.

47,XY,+mar.ish add(17)(p12)(wcp17+,CMT1A+,D17Z1+)
An extra marker chromosome identified by ish to be partially derived from chromosome 17 using whole chromosome paint 17 and probes for *CMT1A* and D17Z1. Additional material of unknown origin has replaced the segment distal to 17p12.

47,XX,+mar.ish der(18)t(18;19)(wcp18+,D18Z1+,wcp19+)
An extra marker chromosome identified by ish to be derived in part from chromosome 18, using whole chromosome paint 18 and an 18 alpha-satellite probe, and from chromosome 19 using whole chromosome paint 19.

13.2.4 Abnormal Mosaic and Chimeric Signal Patterns with Single or Multiple Probes

Cell numbers are given when mosaicism or chimerism is present in constitutional samples.

mos 46,X,+r[15]/45,X[10].ish r(X)(wcpX+,DXZ1+)
A female with two cell lines, one 45,X and another with 46 chromosomes including a minute ring. By ish, the ring was identified as an X using the whole chromosome paint X and the X alpha-satellite probe. The number of cells, not percentage, is given in square brackets. Note that the largest cell line is listed first.

46,X,+r.ish r(X)(wcpX+,DXZ1+)[15]/r(X)(wcpX+,DXZ1++)[10]
A minute ring chromosome replacing a sex chromosome was identified by ish as X using whole chromosome paint X. Probe DXZ1, for the X alpha-satellite, showed the ring to be monocentric in some cells and dicentric in other cells.

mos 45,X[3]/46,XY[12].ish X(DXZ1×1,SRY×0)[32]/X(DXZ1×1),Y(SRY×1)[68]
A male with two cell lines, one 45,X and the other with a normal male karyotype. For determining the sex chromosome complement of an individual, it is appropriate to indicate the status of *SRY* in both cell lines. Using ish, the cell line with a single sex chromosome was confirmed to have one X chromosome but not to contain *SRY*. The 46,XY cell line was confirmed to have a single X chromosome and one appropriately located *SRY* gene on the Y chromosome. A multiplication sign is used, not +, for the ish X as this is a normal signal pattern for the single X chromosome. Note that the normal cell line is listed last.

ish chi X(DXZ1×2)[5]/X(DXZ1×1),Y(SRY×1)[5]
In a chimeric individual, two cell lines, one with two X chromosomes and the other with one X and one Y chromosome, were identified in five metaphases each using probes for DXZ1 and *SRY*.

47,XX,+12[20].ish del(17)(p13p13)(TP53–)[7]/17p13(TP53×2)[13]
A female CLL sample with trisomy 12. FISH with a locus-specific probe detected a *TP53* deletion in 7 of the 20 metaphases examined.

ish del(14)(q21.2q21.2)(RP11-453F20–)dn[6]/14q21.2(RP11-453F20×2)[4]
A *de novo* cryptic deletion within band 14q21.2, originally identified by microarray, was present in 6 of 10 metaphases.

ish del(14)(q21.2q21.2)(RP11-453F20–)dn[16]/dup(14)(q21.2q21.2)(RP11-453F20++) mat[4]
A *de novo* cryptic deletion within band 14q21.2, originally identified by microarray, was present in 16 of 20 metaphases, and a duplication within band 14q21.2, inherited from the mother, was present in 4 of 20 metaphases.

47,XX,+mar[10]/46,XX[20].ish der(15)(:p11.2→q11.2:)(D15S11+,SNRPN+,D15S10–, GABRB3–)[20]/15q11.2q12(D15S11,SNRPN,D15S10,GABRB3)×2[10]
A female mosaic for an additional marker chromosome of unknown origin. FISH identified the marker as chromosome 15 which included the region defined by probes for the *D15S11* and *SNRPN* loci. The level of mosaicism detected was different using FISH.

ish 21q21.3(D21S259/D21S341/D21S342×3)[20]/21q21.3(D21S259/D21S341/D21S342×2)[30]
A mosaic female with three chromosomes 21 present, each with a single copy of the Down syndrome critical region, in 20 of 50 metaphases analysed as determined by ish with a locus-specific probe. All loci of the contig probe are listed in the nomenclature.

13.2.5 Oncology-Specific Exceptions where Multiple Copies of the Same Gene Are Present

An exception to using the multiplication sign can occur in cancer, as shown in these examples below. When the number of signals can be counted, the number of signals should be listed.

ish dmin(MYCN×20~50)[20]
Double minutes, identified to contain *MYCN*, are found in 20–50 copies per cell.

ish ider(21)(q10)dup(21)(q22q22)(RUNX1×4)[5]
An isoderivative chromosome 21 with a duplication of the band q22 is identified by two *RUNX1* copies in each arm of the isoderivative chromosome, for a total of four copies of *RUNX1*.

The abbreviation **amp** can be used if the number of signals cannot be enumerated or meets the clinical definition of amplification for the gene under consideration.

ish ider(21)(q10)add(21)(q11.2)(RUNX1 amp)[3]
An isochromosome derived from chromosome 21 with additional uncharacterized material. It has an increase in *RUNX1* copies meeting the clinical definition of amplification.

ish der(21)(RUNX1 amp)[4]
A derivative chromosome 21 that has an increase in *RUNX1* copies so numerous that they cannot be quantified reliably.

13.2.6 Use of dim and enh in Metaphase *in situ* Hybridization

46,Y,del(X)(p11.4p11.2).ish del(X)(RP11-265P11 dim,RP1-112K5 dim)
Deletion of the short arm of chromosome X. From metaphase FISH on the deleted chromosome, the signals of clones RP1-112K5 (Xp11.2) and RP11-265P11 (Xp11.4) are consistently less intense than on the normal homologue, indicating that they are partially deleted and recognize the proximal and distal breakpoints respectively.

46,XX.ish 17p11.2(RAI1 enh)
Enhanced signal at 17p11.2 by metaphase FISH using a probe to the *RAI1* locus.

13.2.7 Subtelomeric Metaphase *in situ* Hybridization

Subtelomeric FISH is usually performed in panels so that the 41 unique chromosome ends are hybridized simultaneously. A short form is appropriate to describe a normal result after using a subtelomeric FISH panel, for example:

ish subtel(41×2)
Normal result using 41 probes to the 41 subtelomeric regions.

ish der(13)t(13;20)(q34−,p13+)(RP11-63L17−,RP5-1103G7+)
An unbalanced translocation between the distal long arm of chromosome 13 and the distal short arm of chromosome 20. The subtelomeric region of 13q is deleted and replaced with the subtelomeric region of 20p. The designation pter and qter may be used instead of the distal bands. Alternatively, clone names may be used.

ish t(13;20)(q34−,p13+;p13−,q34+)(RP11-63L17−,RP5-1103G7+;RP5-1103G7−,RP11-63L17+)
A balanced translocation between the distal long arm of chromosome 13 and the distal short arm of chromosome 20.

13.3 Interphase/Nuclear *in situ* Hybridization (nuc ish)

Information of interest in *interphase ish*, signified by the symbols **nuc ish**, includes the number of signals and their positions relative to each other. ISCN (1995) provided for the use of a band designation in interphase FISH. This is considered an optional detailed form to be used at the discretion of the investigator or laboratory director. A short form has now been provided that does not indicate chromosome band locations, providing for the ambiguity of hybridization location in interphase nuclei in the absence of discernible bands (chromosomes). Especially in the case of amplification, the short form description is recommended.

When contig probes are used each locus may be listed, separated by **single slant lines (/)** or a single designation may be used in the nomenclature and the composition is described in the report.

13.3.1 Number of Signals

To indicate the number of signals in interphase nuclei, the symbols **nuc ish** are followed immediately in parentheses by the locus designation, a **multiplication sign (×)**, and the number of signals seen. If the detailed form is used, a space should follow **ish**, and then the band designation. For simplicity and readability, the short form is preferred.

If probes for two or more loci are used in the same hybridization, they follow one another in a single set of parentheses, separated by a **comma (,)**, and a **multiplication sign (×)** outside the parentheses if the number of signals for each probe is the same and inside the parentheses if the number of hybridization signals varies.

- If multiple probes on the same chromosome are used, they are listed pter to qter, separated by commas.
- For a single locus visualized with probes to the 3′ and 5′ ends of a gene, the probes should be listed as they reside on the chromosome from pter to qter.
- If loci on two different chromosomes are tested, results are reported in a string, separated by commas, in the order sex chromosomes and autosomes 1 to 22.
- If the study is on a cancer specimen, the number of cells scored is placed in square brackets for each technique.
- Cell lines and clones are listed from largest to smallest number of cells.
- Normal results from multiple hybridizations can be combined in a single set of parentheses; however, if different numbers of cells are studied in multiple hybridizations in a cancer specimen, the results are presented in separate sets of parentheses.
- If two or more techniques are performed, such as chromosome analysis and/or metaphase FISH and interphase FISH, each is reported within the string, separated by a period (.).

Caveats of techniques, for example the semi-quantitative nature of interphase FISH in tissue sections or the design of a particular FISH probe, are not presented in the nomenclature; instead they should be stated in the interpretive text.

13.3.2 Normal Interphase Signal Pattern

nuc ish(KAL1,D21S65)×2
  Two copies of locus *KAL1* and two copies of locus D21S65.

nuc ish(ABL1,BCR)×2[200]
nuc ish 9q34(ABL1×2),22q11.2(BCR×2)[200]
  Two copies of each locus *ABL1* and *BCR* found in 200 cells, expressed with or without band designations.

46,XY[20].ish 9q34(ABL1×2),22q11.2(BCR×2)[20].nuc ish(TP53×2)[200]
  Normal male karyotype in 20 metaphases. A normal hybridization patterns in 20 metaphases using probes for *ABL1* and *BCR* and in 200 nuclei using a probe for *TP53*.

nuc ish(ATM,D12Z3,D13S319,LAMP1,TP53)×2[200]
  Normal hybridization patterns from different hybridizations showing two copies of each of the probes used. Note that they may be listed within one set of parentheses when same numbers of cells are scored.

nuc ish(ATM,TP53)×2[250],(D12Z3,D13S319,LAMP1)×2[200]
  Normal hybridization patterns from different hybridizations showing two copies of each of the probes used. Note that they are listed in separate sets of parentheses when different numbers of cells are scored in separate hybridizations.

46,XY[20].nuc ish(TP53×2)[200]
  Normal male karyotype in 20 metaphases. A normal hybridization pattern in 200 nuclei using a probe for *TP53*.

nuc ish(D17Z1,ERBB2)×2[100]
  Two copies of *ERBB2 (HER-2)* were found in 100 cells with two copies of the centromere 17 probe D17Z1.

nuc ish(D21S65×2)
nuc ish 21q22(D21S65×2)
  Two copies of locus D21S65.

13.3.3 Abnormal Interphase Signal Pattern

When normal and abnormal cells are found, the number of abnormal cells is listed over the total number of cells scored for each abnormal locus. The normal cells are not listed as it is implied that they are the remainder of the total. Probe sets co-hybridized are included in the same parentheses. The probe set with the lowest chromosome number is listed first.

nuc ish(KAL1,GK,DMD)×1
  One copy of each locus, listed pter to qter.

nuc ish(DXZ1×3,SRY×0)
  Three copies of locus DXZ1. The *SRY* signal pattern is given as this is clinically relevant for sex determination.

nuc ish(DXZ1×2,DYZ3×1,D18Z1×3),(RB1,D21S259/D21S341/D21S342)×3
  Three copies of 13, 18 and 21, two copies of X and one copy of Y were found, which may indicate a triploid 69,XXY. Note that the chromosome 21 contig probe shows each locus listed, separated by slant lines.

nuc ish(DXZ1,DYZ3)×1[34/50]/(DXZ1×1,DYZ3×0)[12/50]/(DXZ1×1,DYZ3×2)[6/50]
   One copy of X and one copy of Y in 34 of 50 nuclei, in addition to 12 nuclei with a single X and six nuclei with one X and two Y chromosomes. For determining the sex chromosome complement of an individual, an exception is made to include the normal complement of two sex chromosomes.

nuc ish(TP73×1,ANGPTL×2)[107/200],(ZNF443×2,GLTSCR×1)[105/200]
   Interphase ish shows one *TP73* (maps to 1p36) signal with two *ANGPTL* signals (1q25) in 107 nuclei. In a second hybridization, interphase ish shows two *ZNF443* signals (19p13) with one *GLTSCR* signal (19q13) in 105 interphase nuclei. Thus, the specimen shows loss of both 1p and 19q.

nuc ish(CDKN2C×2,CKS1B×3)[90/100],(FGFR3×2,IGH×3)[93/100],(MYC×2)[100], (CCND1×2,IGH×3)[85/100],(ATM,TP53)×2[100],(IGH×2)(3′IGH sep 5′IGH×1)[95/100], (IGH,MAF)×3(IGH con MAF×2)[87/100],(IGH×3,MAFB×2)[91/100]
   Interphase ish showing gain of 1q (*CKS1B*) and an *IGH-MAF* rearrangement. An extra copy of the *IGH* probe is seen with the IGH/CCND1, IGH/FGFR and IGH/MAFB dual fusion probes. The IGH break-apart probe shows separate 3′*IGH* and 5′*IGH* signals.

nuc ish(ATM,TP53)×2[100],(D12Z3×3,D13S319×2,LAMP1×2)[50/100]
   Two separate hybridizations were performed. In the first, there was a normal hybridization pattern showing two copies each of the probes for the loci *ATM* and *TP53* in all 100 cells. In the second hybridization, there were three signals seen for the probe for the locus D12Z3 and two signals each for the probes for loci D13S319 and *LAMP1* in 50 out of 100 cells.

nuc ish(ATM×1,TP53×2)[100/200],(D12Z3×3,D13S319×2,LAMP1×2)[50/200]
   Two separate hybridizations were performed. In the first, loss of *ATM* signal is found in 100 cells. The remaining cells scored, 100, had a normal signal pattern. In the second hybridization, a gain of signal is seen for D12Z3 in 50 cells. One hundred and fifty cells show the normal pattern. Note, D13S319, *TP53* and *LAMP1* each showed a normal hybridization pattern in 200 cells analyzed.

nuc ish(D13S319×0)[50/200]
   Homozygous deletion of D13S319 in 50 among 200 cells scored. One hundred and fifty cells show a normal pattern.

nuc ish(D13S319×0)[100/200]/(D13S319×1)[50/200]
   Homozygous deletion of D13S319 in 100 among 200 cells scored. Fifty cells show a heterozygous deletion. The 50 remaining cells show a normal pattern. Note that the larger cell line is given first, even though the smaller cell line in this example is likely to represent the stem line.

nuc ish(D13S319×0,LAMP1×2)[100/200]/(D13S319×1,LAMP1×2)[50/200]
   Homozygous deletion of D13S319 in 100 among 200 cells scored. Fifty cells show a heterozygous deletion. The remainder, 50 cells, show a normal pattern. *LAMP1* is used as a control locus and shows two normal hybridization signals in the 200 interphase cells analyzed.

nuc ish(MYCN×12~>50)[200]
   Twelve to more than 50 copies of *MYCN* found in 200 cells.

nuc ish(MYCN amp)[200]
   Number of *MYCN* copies cannot be quantified because it is increased in copy number beyond that which can be counted reliably.

nuc ish(D17Z1,ERBB2)×4~5[100/200]
    Four to five copies of D17Z1 and *ERBB2* were found in 100 out of 200 cells.

nuc ish(D17Z1×2,ERBB2×10~20)[100/200]
    Ten to 20 copies of *ERBB2* were found in 100 cells as compared to two copies of the centromere 17 probe D17Z1.

nuc ish(D17Z1×2~3,ERBB2 amp)[100/200]/(D17Z1,ERBB2)×3[20/200]
    Amplification of *ERBB2* was found in comparison to two to three copies of D17Z1 in 100 cells. In addition, there were three copies of both D17Z1 and *ERBB2* in 20 cells.

nuc ish(D20Z1×2,D20S108×1)[100/200]
    Interphase ish showing one copy of D20S108 as compared to two copies of the centromere probe in 100 cells.

nuc ish(D21S65×3)
nuc ish 21q22(D21S65×3)
    Three copies of locus D21S65.

nuc ish(D21S65,D21S64)×3
    Three copies of locus D21S65 and three copies of locus D21S64.

13.3.4 Donor versus Recipient

Interphase analysis may be used to determine donor versus recipient. For explanation of the use of **double slant line** (//) in chimeras, see section 4.1.

nuc ish(DXZ1×2)[400]//
    400 cells all representing the recipient.

//nuc ish(DXZ1,DYZ3)×1[400]
    400 cells all representing the donor.

nuc ish(DXZ1×2)[50]//(DXZ1,DYZ3)×1[350]
    Fifty recipient XX cells and 350 donor XY cells were found using X and Y centromere probes.

nuc ish(DXZ1×2)[50]//(DXZ1,DYZ3)×1[300]/(DXZ1×1)[10] or
nuc ish(DXZ1×2)[50]/(DXZ1×1)[10]//(DXZ1,DYZ3)×1[300]
    Fifty recipient XX cells were found among 300 donor XY cells and 10 cells with a single X chromosome, listed first as though they are from the donor, and listed secondly as being that they are from the recipient.

13.3.5 Relative Position of Signals

For simplicity and comprehension the short form is preferred.

- If loci on two separate chromosomes are tested, they are expected under normal circumstances to be spatially separated and results are expressed as follows:

nuc ish(ABL1,BCR)×2[100]

● = probe for *ABL1*
○ = probe for *BCR*

- However, if they have become juxtaposed on one chromosome because of a t(9;22), the results are expressed with the first set of parentheses indicating the number of signals and the second set of parentheses describing the relative position of the signals to one another using **sep** and **con**:

nuc ish(ABL1,BCR)×2(ABL1 con BCR×1)[100]

● = probe for *ABL1*
○ = probe for *BCR*

- If they are found to be juxtaposed on two chromosomes using dual-fusion probes, the results are expressed as:

nuc ish(ABL1,BCR)×3(ABL1 con BCR×2)[100]

● = probe for *ABL1*
○ = probe for *BCR*

- If an unusual rearrangement has occurred resulting in the juxtaposition of one *BCR* locus with two *ABL1* loci, the results are expressed as follows:

nuc ish(ABL1,BCR)×3(ABL1 con BCR×1)(ABL1 con BCR con ABL1×1)[100]

● = probe for *ABL1*
○ = probe for *BCR*

- Other possibilities using a dual-fusion probe are as follows:

nuc ish(ABL1,BCR)×2(ABL1 con BCR×1)[100]
  Deletion of the *ABL1-BCR* fusion on one derivative chromosome.

nuc ish(ABL1×2,BCR×3)(ABL1 con BCR×1)[100]
  Deletion of the *ABL1* locus from one fusion on one derivative chromosome.

nuc ish(ABL1,BCR)×4(ABL1 con BCR×3)[198]
  Addition of one *BCR-ABL1* fusion through gain of one derivative chromosome.

- If hybridization signals are normally juxtaposed because of close physical association of the respective loci on the same chromosome, normal results would be expressed as follows:

  nuc ish(KAL1,STS)×2

  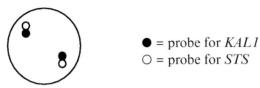

  ● = probe for *KAL1*
  ○ = probe for *STS*

- However, if the loci (example above) are separated because of a structural rearrangement of one X chromosome, the result is expressed as follows:

  nuc ish(KAL1,STS)×2(KAL1 sep STS×1)

  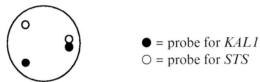

  ● = probe for *KAL1*
  ○ = probe for *STS*

- Amplification of the probe for one of the loci when juxtaposed to a normal signal is expressed as follows:

  nuc ish(IGH×3,BCL2×2,BCL2 amp)(IGH con BCL2×1)(IGH con BCL2 amp×1)[200]

  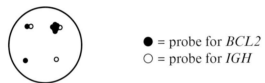

  ● = probe for *BCL2*
  ○ = probe for *IGH*

- Amplification of the probe for one of the loci when separated and not juxtaposed to a normal signal is expressed as follows using **amp**:

  nuc ish(IGH×3,BCL2×2,BCL2 amp)(IGH con BCL2×2)[180]

  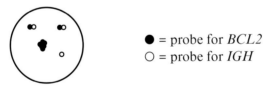

  ● = probe for *BCL2*
  ○ = probe for *IGH*

  nuc ish(PAX7×1,PAX7 amp,FOXO1×1,FOXO1 amp)(PAX7 con FOXO1 amp)[100]
  One normal signal each for the loci *PAX7* and *FOXO1* with fusion of the *PAX7* and *FOXO1* loci on multiple copies of a single chromosome too numerous to count identified through interphase analysis.

  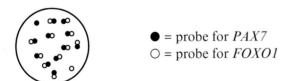

  ● = probe for *PAX7*
  ○ = probe for *FOXO1*

13.3.5.1 Single Fusion Probes

nuc ish(ABL1,BCR)×2(ABL1 con BCR×1)[100]
Single fusion of the *ABL1* and *BCR* loci on a single chromosome identified through interphase analysis.

13.3.5.2 Single Fusion with Extra Signal Probes

nuc ish(ETV6×2,RUNX1×3)(ETV6 con RUNX1×1)[103/200]
ETV6 and *RUNX1* fusion with an extra *RUNX1* signal.

nuc ish(ETV6×1,RUNX1×3)(ETV6 con RUNX1×1)[110/200]
ETV6 and *RUNX1* fusion and deletion of *ETV6* with an extra *RUNX1* signal.

13.3.5.3 Dual Fusion Probes

nuc ish(ABL1,BCR)×3(ABL1 con BCR×2)[200]
Dual fusion of the *ABL1* and *BCR* loci in interphase.

nuc ish(D8Z1×2,MYC×3,IGH×3)(MYC con IGH×2)[100/200],(IGH×3,BCL2×2)[100/200]
Dual fusion of the *MYC* and *IGH* loci in interphase in 100 cells. In a separate hybridization, three copies of *IGH* were seen as compared to two copies of *BCL2* in 100 cells, indicating an *IGH* translocation is present but it does not involve *BCL2*.

nuc ish(MYH11,CBFB)×3(MYH11 con CBFB×2)[100/200]
Dual fusion of the *MYH11* and *CBFB* loci in 100 interphase nuclei.

13.3.5.4 Break-Apart Probes

Given that break-apart probes are made of two probes, the short form does not convey that the normal situation is the presence of two fusion signals. The long form conveys both the normal situation followed by the signal fusion pattern seen.

nuc ish(MLL×2)[200]
nuc ish(5′MLL,3′MLL)×2(5′MLL con 3′MLL×2)[200]
Two MLL probe fusion signals in interphase cells, indicating no disruption of the *KMT2A* gene.

nuc ish(CBFB×2)[200]
nuc ish(5′CBFB,3′CBFB)×2(5′CBFB con 3′CBFB×2)[200]
Two CBFB probe fusion signals in normal interphase cells.

Abnormal cells show the separation of signals.

nuc ish(MLL×2)(5′MLL sep 3′MLL×1)[200]
Two MLL probe signals, but one has separated into the 5′ probe and the 3′ probe, presumably because of a translocation involving the *KMT2A* gene.

nuc ish(5′MLL×2,3′MLL×1)(5′MLL con 3′MLL×1)[213]
   Two 5′MLL probe signals and one 3′MLL probe signal, presumably because of a deletion or because there has been a translocation resulting in 5′*KMT2A*-3′*partner gene*, with loss of the reciprocal derivative chromosome.

nuc ish(3′DDIT3×2,5′DDIT3×1,5′DDIT3 amp)(3′DDIT3 con 5′DDIT3×1)(3′DDIT3 con 5′DDIT3 amp×1)[192/200]
   Using the DDIT3 break-apart probe, a *DDIT3* rearrangement was observed with amplification of the 5′ signal. Note that the orientation is from pter to qter.

● = probe for *5′DDIT3*
○ = probe for *3′DDIT3*

nuc ish(3′IGH×2,5′IGH×3)(3′IGH con 5′IGH×1)[210/237]
   Using the IGH break-apart probe, an *IGH* rearrangement was observed with an extra 5′ signal. Note that the orientation is from pter to qter.

nuc ish(CBFB×2)(5′CBFB sep 3′CBFB×1)[198]
   Two CBFB probe signals, but one has separated into the 5′ probe and 3′ probe, presumably because of an inversion or translocation.

13.3.5.5 Tricolor Probes

nuc ish(GOLIM4×1,MECOM×2,MYNN×2)[55/200]
   A deletion of the proximal region of the long arm of one chromosome 3 including the *GOLIM4* gene region is identified in 55 of 200 interphase nuclei by FISH. There are two MECOM and two MYNN probe signals.

nuc ish(GOLIM4×2,MECOM×3,MYNN×2)(GOLIM4 con MECOM sep MECOM con MYNN×1)[35/100]
   In 35 of 100 nuclei, there is one intact set of fused GOLIM4, MECOM, MYNN probe signals. Due to a rearrangement with a breakpoint within the *MECOM* gene, there is also one GOLIM4 connected to one MECOM signal which is separated from a MECOM signal connected to a MYNN signal.

nuc ish(D4S1036,CHIC2,D4S956)×2[200]
nuc ish(D4S1036,CHIC2,D4S956)×2(D4S1036 con CHIC2 con D4S956×2)[200]
   Two D4S1036, CHIC2, D4S956 probe fusion signals in interphase cells, indicating no disruption of the *CHIC2* gene.

nuc ish(D4S1036×2,CHIC2×1,D4S956×2)(D4S1036 con D4S956×1)[150/200]
   One intact set of D4S1036, CHIC2, D4S956 probe signals, along with one D4S1036 probe signal connected to one D4S956 signal without a CHIC2 signal, due to an interstial deletion including the *CHIC2* gene, in 150 of 200 nuclei.

nuc ish(D4S1036,CHIC2,D4S956)×2(D4S1036 sep CHIC2 con D4S956×1) or (D4S1036 con CHIC2 sep D4S956×1)[150/200]
   One intact set of D4S1036, CHIC2, D4S956 probe signals, along with one D4S1036 probe signal or one D4S956 connected to one CHIC2 signal, due to a rearrangement of the *CHIC2* gene, in 150 of 200 nuclei. The probes for D4S1036 and D4S956 are labeled with the same fluorochrome.

13.4 *In situ* Hybridization on Extended Chromatin/DNA Fibers (fib ish)

Hybridization can be carried out on extended chromatin/DNA fibers usually obtained from interphase nuclei, abbreviated **fib ish**. In this situation, the object of interest is the relative position of the loci at a particular chromosomal location. Where the order of the loci tested is known, they are recorded in the order pter to qter and the chromosomal band is indicated.

■■■ = *D15S11*
●●●● = *SNRPN*
xxxx = *GABRB3*

fib ish 15q11.2(D15S11+,SNRPN+,GABRB3+)
  Signifies that the three loci are present and in the order *D15S11, SNRPN, GABRB3*.

13.5 Reverse *in situ* Hybridization (rev ish)

Reverse in situ hybridization (**rev ish**) refers to the *in situ* hybridization of complex DNA probes derived from a test tissue to normal reference chromosomes. Chromosomes or chromosome segments with *enhanced* (**enh**) or *diminished* (**dim**) fluorescence intensity ratios indicate a relative increase or decrease of the copy number with regard to a basic euploid state. For example, a chromosome present in three copies in a near-diploid cell line would show an enhanced fluorescence intensity ratio, while a chromosome present in three copies in a near-tetraploid cell line would show a diminished ratio. This method can only reveal alterations in copy number of chromosomes or chromosomal segments.

Another method of reverse *in situ* hybridization uses DNA probes derived from parts of the genome from a test tissue, such as the DNA of sorted or microdissected marker chromosomes. *In situ* hybridization of such DNA probes to normal reference chromosomes or to DNA arrays reveals the composition of the isolated chromosome. This method is applicable both to constitutional and acquired abnormalities, and can reveal structural rearrangements not involving copy-number changes (e.g., inversions, balanced translocations).

13.5.1 Chromosome Analyses Using Probes Derived from Sorted or Microdissected Chromosomes

46,XY,add(5)(p15).rev ish der(5)t(5;10)(p15;q22)
  Signifies a derivative chromosome consisting of part of the long arm of chromosome 10 translocated onto the short arm of chromosome 5.

47,XX,+mar.rev ish 15q
  Signifies that the marker chromosome is composed largely or wholly of material from 15q.

13.6 Multi-Color Chromosome Painting

24-color karyotyping and FISH banding are techniques used to paint chromosomes with a distinct color or spectrum of colors. They can be used as a tool to clarify the G-banded analysis. The karyotype can be re-written based on the knowledge gained from the FISH results using these techniques. The use of these FISH techniques should be stated in the report. No special nomenclature has been devised for these techniques. However, a nomenclature similar to that used for **wcp** (Section 13.2) may be used.

13.7 Partial Chromosome Paints

Band-specific or arm-specific probes can be used as *partial chromosome paints* (**pcp**). The nomenclature is similar to that of **wcp** (13.2).

46,XX,?dup(18)(p11.3p11.2).ish dup(18)(pcp18p11.2+)
    A questionable duplication on 18p is shown to contain 18p11.2 material by a partial chromosome paint.

46,XY,inv(8)(p21q13).ish inv(8)(pcp8p++)
    An inversion of chromosome 8 is confirmed by a partial chromosome paint.

14   Microarrays

14.1   Introduction

Microarray-based chromosome analysis is principally an adjunct to traditional chromosome analysis and FISH; for prenatal and postnatal analysis, it has become the laboratory method of choice for definitive identification of chromosome abnormalities. It is increasingly used in cancer studies either as a stand alone test or in combination with FISH. Microarray nomenclature includes the genomic coordinates for banded chromosomes which are defined in the translation tables provided by NCBI (hg19/GRCh38, http://hgdownload.cse.ucsc.edu/goldenPath/hg19/database/cytoBand.txt.gz and hg38/GRCh38, http://hgdownload.cse.ucsc.edu/goldenPath/hg38/database/cytoBand.txt.gz). Platforms differ and evolve; coordinates provided reflect this and are provided only for guidance.

Microarrays can be constructed in at least two ways: with the use of large pieces of cloned DNA such as bacterial artificial chromosomes (BACs), or with small, synthetic sequences of DNA, termed oligonucleotides, which may be designed to detect copy number or to detect single nucleotide polymorphisms (SNPs). Each segment of DNA has a known position within the human genome. These DNA segments are spotted onto a solid support, usually a glass slide or silicon chip, and serve as a target for the genomic DNA sample.

In array-based comparative genomic hybridization (aCGH) using BACs or oligonucleotides, a test DNA and a reference (control) DNA are differentially labeled and simultaneously applied to the microarray. In SNP-based microarray analysis, the patient DNA is hybridized to the microarray and compared by computer analysis to a pool of normal individuals. In either approach, the DNA of the patient is compared to the control or reference DNA and gains or losses can be detected. The ISCN nomenclature is appropriate for either type of array data and for arrays that contain combinations of oligonucleotides and SNPs. In this nomenclature, the number and type of clones used as targets (BAC, cosmid, fosmid, oligonucleotide, etc.) are not included.

Two systems have been devised; a detailed form that includes the span of the abnormal nucleotides as well as the bordering normal nucleotides, and a short form that includes only the abnormal nucleotides. In the short form the nucleotide numbers are given either with or without commas to indicate thousands and millions. In the detailed form the nucleotide numbers are given without the demarcating commas. The span of affected nucleotides is separated by an **underscore (_)**, in line with the Human Genome Variation Society (HGVS) recommendations (www.HGVS.org/varnomen) for molecular genetic nomenclature. It is acceptable to use a mixture of detailed form and short form to describe different abnormalities within a cytogenomic profile.

If microarray clarifies a karyotype and, in retrospect, the abnormality can be visualized with banding, the karyotype may be re-written to reflect this new microarray information. If the abnormality is cryptic and cannot be visualized by banding, the abnormality should not be listed in the banded karyotype. Copy number gains and losses are not indicative of chromosome architecture and so other methods, e.g., chromosome banding, FISH or long read sequencing, are needed for confirmation.

In highly complex array results, as may be the case in cancer studies, the laboratory may choose to display results using ISCN nomenclature in table form instead of in a string. The information in the table must include the chromosomes and bands corresponding to the variant, the type of variant (loss, gain, amplification or region of homozygosity), the designated genome build and the genomic coordinates of the variant. Mosaicism must be indicated if present. It is recommended that the copy number and an estimate of the proportion of the sample with the abnormality are reported. Alternatively mosaicism can be indicated by a range of copy number. At the discretion of the laboratory, the size of the variant (in kb or Mb) and relevant genes within the affected region may also be included in the table.

Copy number variation can also be identified by next-generation sequencing (NGS) technologies, also known as massively parallel sequencing (MPS). In this method whole genome, whole exome or targeted sequence data are analyzed using CNV calling software. This technique is further discussed in Chapter 16.

14.2 Examples of Microarray Nomenclature

14.2.1 Normal

If *no abnormality* is detected using an array that has probes targeted to multiple loci across all chromosomes, the results are expressed as follows with no space between **arr** and the opening parenthesis. The sex chromosomes are listed before the autosomes.

arr(X,1–22)×2   normal female
arr(X,Y)×1,(1–22)×2   normal male

The descriptive narrative, or interpretation, in the report should indicate the platform used, the resolution, and whether the array represents the entire genome of all chromosomes.

14.2.2 Abnormal

If the results are *abnormal*, list only the aberrations. Regardless of whether there is a copy number gain or loss, the aberrations of sex chromosomes are listed first followed by autosomes which are listed from lowest to highest number chromosome. The band designations of only the abnormal genomic regions are shown. The aberrant nucleotides are listed from pter to qter, consistent with the public databases of current genome builds on UCSC or Ensembl Genome Browsers (www.genome.ucsc.edu or www.ensembl.org). An underscore is used to indicate that the gain or loss encompasses the segment between the listed nucleotides. When nucleotide coordinates are used to define an abnormal result, the specified genome build (e.g., [GRCh38] referring to the Genome Reference Consortium Human Build 38 assembly) is placed within the string as illustrated in this chapter. Note that there is a space after the square bracket. The specific genome build is not necessary when describing a normal male or female result or an aneuploidy with the short description as shown in this chapter.

arr(X)×2,(Y)×1
  Microarray analysis shows a single copy gain of the X chromosome in a male.

arr(X)×2,(Y)×1,(1–22)×3
  Microarray analysis shows triploidy 69,XXY.

arr(X,1–22)×3
  Microarray analysis shows triploidy 69,XXX.

arr(X,1–19,21,22)×3
  Microarray analysis shows a near-triploid female with only two copies of chromosome 20.

arr[GRCh38] Xq28 or Yq12(155790872_156030895 or 56959165_57217415)×1
  Microarray analysis shows a single copy terminal loss of the pseudoautosomal region that normally is found at Xq28 and Yq12. It is not possible to determine if the loss is from X or Y in a male; FISH or chromosome analysis is required to confirm the origin of the loss.

arr[GRCh37] (X)×2,Yp11.31(2,650,140_2,841,956)×1
  Microarray analysis shows a single copy gain of Yp11.31 chromosome material with no discernable loss from two X chromosomes nor from autosomes. FISH or chromosome analysis is required to confirm the structural nature of the gain which includes the *SRY* gene.

arr[GRCh37] Yq11.23(26887746_27019505)×0,20q13.2q13.33(51840606_62375085)×3
  Microarray analysis shows an interstitial loss of the proximal long arm of the Y chromosome and a gain of the distal long arm of chromosome 20. Note that the sex chromosome abnormality is listed first.

arr[GRCh38] 1p36.33p36.32(827048_3736354)×3,1q41q44(221649655_247175095)×1
  Microarray analysis shows a terminal gain of the short arm of chromosome 1 and a terminal loss of the long arm of chromosome 1. This result may indicate a duplication/deletion recombinant chromosome from an inversion parent, but further studies of the parents and/or child by FISH or chromosome analysis are required.

arr[GRCh38] 4q32.2q35.1(163,146,681_183,022,312)×1
or
arr[GRCh38] 4q32.2q35.1(163002425×2,163146681_183022312×1,184322231×2)
  Microarray analysis shows an interstitial loss of the long arm of chromosome 4 at bands q32.2 through q35.1, which is at least 19.9 Mb in size. The detailed form shows that the next neighboring proximal nucleotide that does not show a loss is 144,256 nucleotides away and the next neighboring distal nucleotide that does not show a loss is 1.3 Mb away from the alteration. Note that delimiting commas are not used in the detailed description.

arr[GRCh38] 6q21q25.1(113,900,000_149,100,000)×1,(21)×3
  Microarray analysis (based on the GRCh38 assembly) shows interstitial loss in the long arm of chromosome 6 at bands q21 through q25.1 and a single copy gain (trisomy) of chromosome 21.

arr(8,21)×3
  Microarray analysis shows a single copy gain of chromosomes 8 and 21.

arr[GRCh38] 9p24.3p13.1(204166_38756057)×1,18q21.33q22.1(63877984_64683663)×1,
21q11.2q21.1(13600026_20175986)×3

    Microarray analysis shows three abnormalities; a terminal deletion of the short arm of the 9p covered by the array, an interstitial deletion of the long arm of chromosome 18 and a gain of the distal long arm of chromosome 21. Note that the chromosomes are listed in numerical order, regardless of whether they show a gain or loss.

arr[GRCh38] 11p12(37741458_39209912)×3

or

arr[GRCh38] 11p12(37003221×2,37741458_39209912×3,39752007×2)

    Microarray analysis shows a single copy gain of the short arm of chromosome 11 at band p12. The gain is at least 1.47 Mb in size. The next neighboring distal nucleotide that does not show a gain is 738,237 nucleotides away from the alteration and the next neighboring proximal nucleotide that does not show a gain is 542 kb away from the alteration.

arr[GRCh38] 14q31.1(82695844_82855387)×1,14q32.33(105643093_106109395)×3

    Microarray analysis shows two abnormalities on chromosome 14. Note that the abnormalities are shown from pter to qter, irrespective of whether they are gains or losses.

arr[GRCh38] 15q11.1q13.2(20,366,669_30,226,235)×4

    Microarray analysis shows a two copy gain of proximal 15q, resulting in tetrasomy 15q11.1q13.2. Distinguishing a supernumerary marker chromosome or an interchromosomal insertion from a triplication of this region requires FISH.

arr(16q)×3

    Microarray analysis shows a single copy gain of the whole long arm of chromosome 16.

arr[GRCh38] 18p11.32q23(102328_79093443)×3

or

arr(18)×3

    Microarray analysis shows a single copy gain of the entire chromosome 18, consistent with trisomy 18.

arr[GRCh38] 18p11.32p11.21(102328_15079388)×1,18q22.3q23(69172132_79093443)×1

    Microarray analysis shows a single copy terminal loss of the distal short arm of chromosome 18 and a single copy terminal loss of the distal long arm of chromosome 18, likely indicating a ring chromosome 18, although FISH or chromosome analysis is required to confirm.

arr[GRCh38] 20q13.13q13.33(51001876_62375085)×1,22q13.32q13.33(48533211_49525263)×3

    Microarray analysis shows a single copy interstitial loss of 20q and a single copy gain of 22q.

arr[GRCh38] 21q11.2q22.3(13531865_46914745)×3

    Microarray analysis shows a single copy gain of the entire long arm of chromosome 21, likely indicating trisomy 21. Note that most microarrays will not have coverage of the repetitive short arm sequences; thus the short arm is not designated. Trisomy 21 is implied, but FISH or chromosome analysis is required to exclude a Robertsonian or other translocation.

14.2.3 Inheritance

Microarray analysis can demonstrate only a relative gain or loss of DNA; thus, FISH analysis or karyotype is necessary to demonstrate the structure of deletions, duplications, insertions, unbalanced translocations, etc. definitively. The parental origin of the abnormality may follow the copy number (×1, ×3, etc.). There is a space between the copy number and the inheritance symbol (**dn, mat, pat, inh, dmat, dpat, dinh**) but no space if the inheritance symbol follows a parenthesis in the detailed form.

arr[GRCh38] Xq25(126228413_126535347)×0 mat
or
arr[GRCh38] Xq25(126023321×1,126228413_126535347×0,126556900×1)mat
 Microarray analysis shows an interstitial loss of the long arm of the X chromosome at band q25 in a male. The hemizygous loss is at least 306,934 nucleotides. The next neighboring proximal nucleotide that does not show a loss is 205,092 nucleotides away, and the next neighboring distal nucleotide that does not show a loss is 21,553 nucleotides away from the alteration. This deletion was inherited from the mother.

arr[GRCh38] Xq25(126,228,413_126,535,347)×1 mat
or
arr[GRCh38] Xq25(126023321×2,126228413_126535347×1,126556900×2)mat
 Same abnormality as the above example, but found in a female.

arr[GRCh38] Xp22.31(6467202_8091950)×0 mat
 Microarray analysis in a male shows an interstitial loss of the short arm of the X chromosome at band p22.31, inherited from a carrier mother.

arr[GRCh38] Xp11.22(53215290_53986534)×2 mat
 Microarray analysis in a male shows a gain of the short arm of the X chromosome at band p11.22, inherited from a carrier mother.

arr[GRCh38] Xp11.22(53215290_53986534)×3 mat
 Same abnormality as the above example, but found in a female.

arr[GRCh38] 4q28.2(128184801_129319376)×3 mat,16p11.2(29581254_30066186)×3 pat
 Microarray analysis shows a gain of 4q, inherited from the mother, and a gain of 16p, inherited from the father.

arr[GRCh38] 4q32.2q35.1(163146681_183022312)×1 dn
or
arr[GRCh38] 4q32.2q35.1(163002425×2,163146681_183022312×1,184322231×2)dn
 Microarray analysis shows an interstitial loss of the long arm of chromosome 4 between bands q32.2 and q35.1. The heterozygous loss is *de novo*.

arr[GRCh38] 9p24.3(1310386_1709409)×1 mat,9p22.3p22.2(16455330_16763471)×1 dn, 18q21.33q22.1(62747805_67920791)×1 dn
 Microarray analysis shows three abnormalities. The first abnormality is an interstitial loss of maternal origin; the other two abnormalities are *de novo* interstitial losses. Therefore, the inheritance of each is listed after the specific gain or loss. Note that the two abnormalities on chromosome 9 are listed from pter to qter.

arr[GRCh38] 17p11.2(16512256_20405113)×3 dn
>Microarray analysis shows a single copy gain of the short arm of chromosome 17 at band p11.2. The gain is at least 3.89 Mb in size and is *de novo* in origin.

arr[GRCh37] 22q11.21(18,916,842_21,465,659)×4 mat pat
>Microarray analysis shows a gain of two copies of 22q; one gain is inherited from the mother and the other is inherited from the father. Each parent carries three copies and passes on two copies to their offspring, i.e., they have inherited the whole abnormality from each parent.

14.2.4 Multiple Techniques

Observations combined with banded chromosome analysis and microarrays can be expressed by using the symbol **arr** followed by the genome build, a space and the chromosome region, band or sub-band designation of the locus. A period (.) precedes the microarray nomenclature. Where multiple techniques are used, the karyotype is listed first, followed by FISH, followed by results obtained with other techniques. Alternatively, results obtained using different techniques may be presented on separate lines without periods.

Note that the conventional cytogenetic banding assignments are those derived from banded chromosomes, while the array banding assignments are those derived from genome browsers. These are not always concordant.

46,X,der(Y)t(Y;20)(q11.23;q13.2).arr[GRCh38] Yq11.23(26887746_27019505)×0, 20q13.2q13.33(51840606_62375085)×3
>Microarray analysis shows a terminal loss of the long arm of the Y chromosome and a terminal gain of the long arm of chromosome 20 which banded chromosomes show are due to an unbalanced translocation between the long arm of the Y chromosome and the long arm of one chromosome 20. Note that the array nomenclature lists the sex chromosome abnormality first. There is no normal Y chromosome in this individual.

46,XY,der(20)t(Y;20)(q11.23;q13.2).arr[GRCh37] Yq11.23(26887746_27019505)×2, 20q13.2q13.33(51840606_62375085)×1
>Microarray analysis shows an unbalanced translocation derived from a rearrangement between the long arm of the Y chromosome and the long arm of one chromosome 20, resulting in a terminal gain of distal Yq and a terminal deletion of distal 20q. Note that the array nomenclature lists the sex chromosome abnormality first and that there is a normal Y chromosome in addition to the derivative chromosome 20 in this individual.

46,XX.arr[GRCh38] Xp22.31(6923924_7253485)×3,5q14.3(88018766_89063989)×1
>Normal female karyotype with microarray analysis shows a single copy gain of part of the short arm of the X chromosome and a single interstitial copy loss of part of the long arm of chromosome 5. Note that the sex chromosome abnormality is listed first.

46,X,der(Y)t(X;Y)(p22.33;q12).arr[GRCh37] Xp22.33(701_2,679,502)×3,Xp22.33p22.2 (2,709,521_15,955,588)×2,Yq11.221q11.23(16,139,805_27,177,529)×0
>Karyotype and microarray analyses show a single copy gain of the X chromosome from two regions of Xp and loss of the long arm of the Y chromosome, resulting from an unbalanced translocation between the short arm of the X chromosome and the long arm of the Y distal to Yq11.221. The two regions of Xp are shown separately because the gain of the pseudoautosomal region results in three total copies and the gain proximal to the pseudoautosomal region results in two total copies.

ish der(X)t(X;Y)(p22.33;p11.2)(SRY+).arr[GRCh37] Xp22.33(168,546_2,664,272)×1, Xp22.33q28(2,664,962_155,232,907)×2,Yp11.32p11.2(11,091_3,163,644)×1,Yp11.2q11.23 (3,713,948_2,872,2435)×0

> Microarray analysis is interpreted in the context of metaphase FISH which identified an unbalanced translocation derived from rearrangement between the short arms of the X and Y chromosomes, resulting in a loss of distal Xp and most of the Y chromosome. The translocated portion of Yp including PAR1 and the region harbouring *SRY* can be attributed to the Y chromosome.

47,XY,+mar.arr[GRCh37] 1p13.1p11.2(117596421_121346616)×3 dn

> Microarray analysis shows a single copy gain of the short arm of chromosome 1, spanning approximately 3.4 Mb, likely identifying the marker chromosome. Because most microarrays will not contain the heterochromatin near the centromeres, the centromeric bands are rarely included in the nomenclature of rings and markers after microarray analysis, although the centromere is probably included in the aberration and would need to be confirmed by FISH. An amended result after FISH analysis could be written as:

47,XY,+mar.ish r(1)(p13.1q1?1)(D1Z1+).arr[GRCh37] 1p13.1p11.2(117596421_121346616)×3 dn
or
47,XY,+mar
ish r(1)(p13.1q1?1)(D1Z1+)
arr[GRCh37] 1p13.1p11.2(117596421_121346616)×3 dn

47,XY,+mar.arr[GRCh38] 1p12p11.2(117596421_121013236)×3 dn,15q25.1q26.3 (78932946_100201136)×3 dn

> Microarray analysis shows two *de novo* aberrations: a single copy gain of part of the short arm of chromosome 1 and a single copy gain of part of the long arm of chromosome 15. This likely identifies a complex marker comprised of material from chromosome 1 and chromosome 15, but FISH would be required for confirmation.

46,XX.arr[GRCh38] 3p12.2(80395073_83498191)×3 inh,12p12.1(23543231_23699047)×1 dn

> Normal female karyotype analysis showing a gain of chromosome 3 at band 3p12.2 by microarray analysis, inherited from a parent, and an interstial loss of 12p at band 12p12.1 of *de novo* origin.

arr[GRCh38] 8q23.1q24.3(105171556_146201911)×3,15q26.2q26.3(96062102_100201136)×1

> Microarray analysis shows a single terminal copy gain of part of 8q and a single terminal copy loss of part of 15q. Often, double segmental imbalances are indicative of unbalanced translocations. However, microarrays can only detect relative imbalances in DNA copy number. After chromosome analysis and FISH visualization, the nomenclature can be written as follows:

46,XY,der(15)t(8;15)(q22.3;q26.2)dmat.ish der(15)t(8;15)(RP11-1143I12+,RP11-14C10–).arr[GRCh38] 8q23.1q24.3(105171556_146201911)×3,15q26.2q26.3 (96062102_100201136)×1
or
46,XY,der(15)t(8;15)(q22.3;q26.2)dmat
ish der(15)t(8;15)(RP11-1143I12+,RP11-14C10–)
arr[GRCh38] 8q23.1q24.3(105171556_146201911)×3,15q26.2q26.3(96062102_100201136)×1

> **dmat** is placed after the conventional cytogenetic result because the derivative was determined to be inherited from a balanced translocation in the mother.

46,XY,rec(18)dup(18q)inv(18)(p11.32q21)dpat.arr[GRCh38] 18p11.32(102328_
2326882)×1,18q21.31q23(56296522_76093443)×3
> Chromosome analysis shows an abnormal chromosome 18 which is interpreted in the context of microarray and paternal karyotype. Microarray analysis shows a terminal loss of the short arm of chromosome 18 and a gain of the terminal region of the long arm of chromosome 18. Karyotype analysis in the father of this individual demonstrated a balanced pericentric inversion. Thus, this is a duplication/deletion recombinant chromosome that is derived from an inversion carrier parent.

47,XX,+mar.arr[GRCh38] 21q11.2q21.1(13461349_17308947)×4,21q22.3(46222759_
46914885)×3
> Microarray analysis shows a two copy gain of 21q11.2q21.1 and a single copy gain of 21q22.3, indicating that the marker chromosome is likely a complex rearrangement involving two different segments of chromosome 21, resulting in partial tetrasomy of proximal 21q and partial trisomy of distal 21q.

47,XX,+mar[11/20].arr[GRCh38] 21q11.2q22.3(13461349_46914885)×3
> Microarray analysis of a cancer sample shows a gain of 21q11.2q22.3, indicating that the marker chromosome is likely a complex rearrangement involving chromosome 21.

14.2.5 Mixed Cell Populations and Uncertain Copy Number

To indicate a mixed cell population, the proportion of the sample with the abnormality can be estimated and included in brackets following the copy number. If the proportion of abnormal DNA cannot be estimated, the copy number range should be given using a tilde (~) or **mos** may be used.

Note that the gains and losses are reported relative to the normalised ploidy and that it may not be possible to distinguish between a one copy gain in a high proportion of the sample and a two copy gain in a low proportion of the sample. Similarly it may be necessary to use a different method to determine the ploidy of a sample, e.g., to discern tetraploidy from diploidy.

arr(X)×1[0.6]
or
arr[GRCh38] Xp22.33q28(168,546_155,233,730)×1[0.6]
> Microarray analysis shows a single copy loss of the X chromosome in approximately 60% of the sample in a female.

arr[GRCh38] Xp22.33q28(168,546_155,233,730)×1~2
or
arr[GRCh38] Xp22.33q28(168,546_155,233,730)×1 mos
> Microarray analysis of a female shows copy number loss of the X chromosome in a proportion of cells which is not determined.

arr[GRCh37] Xp22.33q26.2(61090_132353367)×1[0.8],Xq26.2q28(132435273_
151904036)×1
> In a female, microarray analysis shows that 100% of cells are missing one copy of Xq26.2 to Xq28: 80% have one normal X chromosome only, while 20% contain a normal X and an X chromosome with a terminal deletion of Xq26.2 to qter.

arr[GRCh38] Xp22.33p11.23(701_48,643,784)×1[0.75],Xp11.23q21.1(48,643,785_
77,173,852)×1[0.25],Xq21.1q28(77,173,853_155,270,560)×1[0.75]
> In a female, microarray analysis shows mosaicism for X chromosome losses with the segment from Xp11.23 to Xq21.1 exhibiting a lower level of abnormality than the Xpter and Xqter segments, suggestive of mosaic ring chromosome X. FISH or chromosome analysis is required to determine the structural nature of the abnormalities.

arr[GRCh38] (X,1–7)×1,(9–12)×1,13q12.11q14.2(19438807_48800573)×1,13q14.2
(48986461_49176936)×0~1,13q14.2q34(49176999_115095706)×1,(15–20)×1,(22)×1

    Microarray analysis in a cancer in a female shows a near haploid genome, with single copies of chromosomes 1 to 7, 9 to 12, 15 to 20, 22 and X. The result for chromosome 13 shows regions of single copy loss and a region in 13q14.2 with a mixture of homozygous loss and single copy loss. Note that microarray cannot determine whether the two mono-allelic deletions occur in cis or in trans. Chromosomes 8, 14, and 21 show the normal two copies.

ish mos del(2)(q11.2q13)(RP11-478D22–)[10]/2q12.1(RP11-478D22+)[25].
arr[GRCh38] 2q11.2q13(100982729_112106760)×1[0.4]

    FISH and microarray analyses show a mosaic deletion in the long arm of chromosome 2. By microarray approximately 40% of cells have the deletion.

47,XY,+mar.ish der(2)(p11.2q13)(RP11-478D22+)[5]/2q12.1(RP11-478D22+)[25].
arr[GRCh38] 2p11.2q13(90982729_112106760)×3[0.15]

    Microarray and FISH analyses demonstrate a mosaic marker chromosome derived from chromosome 2. By microarray approximately 15% of cells have 3 copies of the defined region.

arr[GRCh37] (5,6)×3[0.3],7q34(138588953_140233585)×2~>2

    SNP microarray analysis of a solid tumor shows three copies of chromosomes 5 and 6 in 30% of the sample. There is mosaic gain of a region within 7q34, but the copy number is uncertain. There is allelic imbalance demonstrated in the B allele frequency plot for this region, but it is not possible to determine the exact copy number. Where clinically relevant, copy number should be confirmed by FISH.

arr[GRCh38] 7p11.2(54290345_55087100)amp

    Microarray analysis of a solid tumor shows amplification of a region in 7p11.2. The exact copy number is too high to be enumerated accurately by array.

arr[GRCh38] 11q22.3q23.2(104669588_113439979)×1[0.3],13q14.13q14.3(46290874_51390298)×1[0.8]

    Microarray analysis of DNA from a CLL patient shows deletion in the long arm of chromosome 11 in approximately 30% along with a deletion in the long arm of chromosome 13 in approximately 80% of the sample.

arr[GRCh38] 12p13.33p11.1(84917_34382567)×2~4

    Microarray analysis shows a two copy gain of the short arm of chromosome 12, resulting in tetrasomy 12p. Although this result likely indicates an isochromosome of 12p, such as those found in Pallister Killian syndrome, FISH or chromosome analysis is required to confirm. The approximate sign is used to indicate that the number of copies of this region varies from 2 to 4.

arr[GRCh37] 13q14.2(50,487,993_50,512,864)×1[0.9],13q14.2q14.3(50,531,767_51,375,971)×0[0.9],13q14.3(51,379,765_51,711,436)×1[0.9]
or
arr[GRCh37] 13q14.2(50,487,993_50,512,864)×1~2,13q14.2q14.3(50,531,767_51,375,971)×0~2,13q14.3(51,379,765_51,711,436)×1~2

    The oncology microarray profile shows three contiguous deletions on the long arm of chromosome 13 resulting in a region of homozygous loss with regions of heterozygous loss on either side. Note that microarray cannot determine whether the two heterozygous deletions occur in cis or in trans.

14.2.6 Nomenclature Specific to SNP Array

Single nucleotide polymorphisms (SNP) can be identified using certain types of oligonucleotide microarrays. The use of SNP arrays may uncover regions of homozygosity that have been reduced from previously known heterozygosity. The symbols **htz** and **hmz** can be used to define the zygosity of the chromosomal region. SNP arrays are also used in the detection of abnormalities relative to genome ploidy.

arr[GRCh38] 11p12(37741458_39209912)×2 hmz
SNP array analysis shows homozygosity in the short arm of chromosome 11, at band p12, at least 1.47 Mb in size.

arr[GRCh38] 11p12(37741458_39209912)×2 hmz mat[0.6]
SNP array analysis shows mosaic homozygosity in the short arm of chromosome 11, at band p12, at least 1.47 Mb in size in approximately 60% of the sample and of maternal origin.

arr[GRCh38] 15q11.2q26.3(23123715_101888908)×2 hmz pat,21q11.21q22.3(14595263_48084819)×2 hmz pat
SNP array analysis shows homozygosity for the entire long arms of chromosomes 15 and 21, respectively. Based on additional SNP array analyses in the parents, both regions of homozygosity reveal uniparental isodisomy obtained from the father.

arr(15q,21q)×2 hmz pat
or
arr[GRCh38] (15q11.2q26.3(23123715_101888908),21q11.21q22.3(14595263_48084819))×2 hmz pat
Same example as above. In some circumstances, a short form may be sufficient to describe the abnormalities. Note the use of the parentheses in the detailed form with multiple regions of homozygosity being grouped.

arr(7)×2 htz mat
SNP array analysis shows maternal uniparental heterodisomy for chromosome 7.

arr[GRCh38] 11p15.5p15.4(2265338_6275434)×2 hmz c,19q13.33q13.43(49759500_58586384)×2 hmz
This is a possible example of a Beckwith-Wiedemann syndrome patient with constitutional segmental UPD for 11p15.5p15.4 and an acquired region of homozygosity of 19q13.33qter. Segmental UPD may be better referred to in cancer cases as copy neutral loss of heterozygosity (LOH) and in constitutional cases as absence of heterozygosity (AOH).

arr[GRCh38] 11p15.5p15.4(2265338_6275434)×2 hmz,19q13.33q13.43(49759500_58586384)×2 hmz
In this second example, both regions of homozygosity are acquired or the inheritance is not known.

arr[GRCh37] (X,3,7,9q,13–17,19,20,22)×1
SNP microarray profile result indicating near-haploidy in a female with acute lymphoblastic leukemia. Chromosomes 1, 2, 4–6, 8, 9p, 10–12, 18, and 21 show a heterozygous state, which does not need to be emphasized, since this represents the normal situation, whereas the other chromosomes and 9q show homozygosity, which also does not to be emphasized because this is already clear by the ×1 notation. Based on only SNP microarray data, it is not always possible to conclude/deduce whether this concerns a near-haploid complement or a mixture of near-haploid and doubled near-haploid complements.

To describe the above as a doubled clone: arr[GRCh37] <2n>(X)×2 hmz,(1,2)×4,(3)×2 hmz,(4–6)×4,(7)×2 hmz,(8,9p)×4,(9q)×2 hmz,(10–12)×4,(13–17)×2 hmz,(18)×4,(19,20)×2 hmz,(21)×4,(22)×2 hmz

In this way, it becomes clear that all chromosomes are actually abnormal compared to normal diploid, either by copy number or by being homozygous.

arr(X,Y)×1~2,(1–13)×2 mos hmz,(14)×2~4,(16–20)×2 mos hmz,(21)×2~4,(22)×2 mos hmz
 Consistent with a doubled near-haploid complement in a male with acute lymphoblastic leukemia. There is a mixture of DNA from normal and leukemic cells; thus, SNP array analysis shows mosaic homozygosity for chromosomes 1 to 13, 16 to 20, and 22. There was copy number gain for chromosomes X, Y, 14, and 21.

arr[GRCh37] 12q11~q13.13(37,876,400_51,566,350)×2 hmz[0.2~0.3],12q13.13q24.33 (51,566,541_133,777,645)×2 hmz[0.3]
or
arr[GRCh37] 12q11~q13.13(37,876,400_51,566,350)×2 mos hmz,12q13.13q24.33 (51,566,541_133,777,645)×2 mos hmz
 By SNP microarray the region of chromosome 12 between q11 and q13.13 displays mosaic copy neutral loss of heterozygosity (CN-LOH), and there is gradual increase in separation of the B allele frequency heterozygous plot indicating that there are numerous subclones with slightly differing breakpoints. The tilde (~) between the chromosomal breakpoints indicates a range to demonstrate these multiple subclone breakpoints. The mosaicism of this region, as calculated from the BAF, ranges from approximately 20 to 30% of the sample. The contiguous region from 12q13.13 to q24.33 displays mosaic CN-LOH of approximately 30%, with no discernible subclonal variation.

arr[GRCh37] 17p13.3~p13.1(8,547_7,184,396)×1[0.7~0.9],17p13.1p11.2(7,184,481_17,446,103)×1[0.9]
 SNP microarray demonstrates loss of 17p with a range of deletion breakpoints between 17p13.3 and 17p13.1, as indicated by the use of "~" in the break point designation. There is similarly change in the level of mosaicism across this region from 70 to 90% of the sample. Deletion of the region between 17p13.1 and 17p11.2 appears as a single block with no discernible subclonal variation.

arr[GRCh37] 17p13.3~p13.1(8,547_7,184,396)×1~2,17p13.1p11.2(7,184,481_17,446,103)×1~2
 The same microarray profile as above, but the proportion of the sample bearing the abnormalities was not determined.

arr[GRCh37] 13q14.13q14.2(46389795_50424677)×2 hmz[0.40],13q14.2q14.3(50438236_51692174)×0[0.4],13q14.3q34(51692548_115103529)×2 hmz[0.40]
 SNP microarray demonstrates 40% mosaic homozygosity for most of the long arm of chromosome 13. There is a deletion within the homozygous region that is therefore a biallelic deletion.

arr[GRCh37] 13q14.13q14.2(46389795_50424677)×2 mos hmz,13q14.2q14.3(50438236_51692174)×0~2,13q14.3q34(51692548_115103529)×2 mos hmz
 The same microarray profile as above, but the level of mosaicism was not determined.

arr<3n>(X)×2,(Y)×1
 A triploid cytogenomic profile with disomy X and one Y chromosome.

arr<4n>(7)×3[0.9]
 Loss of one copy of chromosome 7, relative to the tetraploid genome, in 90% of the sample.

arr<2n>(X,4,6,8,10,11)×3[0.3],(14)×4[0.3],(17,18)×3[0.3],(21)×4[0.3]
    A near-triploid cytogenomic profile for an ALL sample with the banded karyotype: 58<2n>,XY,+X,+4,+6,+8,+10,+11,+14,+14,+17,+18,+21,+21[5]/46,XY[15]. Since it is biologically relevant in this case the abnormal array profile has been described relative to the diploid genome.

14.2.7 Complex Array Results

The symbol **cx** for complex chromosome rearrangement is used for multiple complex rearrangements across the entire genome or within a region of the genome.

arr(X,1–22)cx
    Microarray analysis shows multiple complex rearrangements across the entire genome in a female.

arr(1–22)cx
    Microarray analysis shows multiple complex rearrangements in chromosomes 1 through 22. The sex chromosomes appear normal and are therefore not shown.

arr(X,Y,1–22)cx
    Microarray analysis shows multiple complex rearrangements across the entire genome in a male.

arr[GRCh37] 3p26.3q12.1(61495_98386666)cx[0.5]
    Microarray shows a complex pattern of chromosomal copy number changes in the short arm of chromosome 3. It affects approximately 50% of the sample.

Chromothripsis (**cth**) refers to complex patterns of alternating copy number changes (commonly alternating disomy and heterozygous loss) clustered along a chromosome or chromosomal segment.

arr(1p)cth
    Microarray analysis shows multiple alternating changes (normal segments, gains, and/or losses within the region) in the short arm of chromosome 1. Only the short arm of chromosome 1 is affected.

arr(1,13)cth
    Short description of a microarray analysis that shows chromothripsis in chromosome 1 and chromosome 13. Only chromosomes 1 and 13 are affected.

arr[GRCh38] (1)cth,6q25.1q27(149100000_170899992)×1,(13)cth
    Microarray analysis shows chromothripsis as in the above example and concomitant loss of the long arm of chromosome 6 at bands q25.1 through q27.

arr[GRCh37] 2p24.3p21(13197725_46386298)cth[0.9]
    Microarray shows chromothripsis ocurring within the region 2p24.3 to 2p21 in approximately 90% of the sample.

Chromoanasynthesis (**cha**) refers to a complex pattern of copy number changes (commonly deletions and one and two copy gains) which occur as a result of DNA replication machinery defects and which affect a single chromosome or chromosome region. The chromosomal pattern includes copy number variants without the clustered breakpoints of chromothripsis.

arr(6)cha
    Microarray analysis shows multiple changes of copy number, between 1 and ~4, affecting chromosome 6.

arr 17p13.3q11.2(1,207,467_28,236,645)cha[0.5]
: Microarray analysis shows multiple changes of copy number, between 1 and ~4, affecting most of the short arm of chromosome 17. Chromoanasynthesis was evident in approximately 50% of the sample.

14.2.8 Polar Bodies

First and second polar bodies as well as the secondary and the fertilized oocyte are the result of meiotic division (reduction division) and normally do not contain chromosomes, but one or two chromatids (the same is true for secondary and tertiary spermatozoa). This section describes nomenclature for chromatid loss and gain in polar bodies or respective oocytes identified by the cytogenomic techniques of microarray and shallow next-generation sequencing (shallow NGS).

Following meiosis 1, the first polar body (PB1) consists of two chromatids, usually from a single chromosome for 1–22 and the X.

At fertilization, meiosis 2 is completed forming the second polar body (PB2) which consists of only one chromatid for 1–22 and X.

When describing normal or abnormal PB results, the term **cht** (chromatid) is used. For PB1 it is relative to a normal haploid set of two chromatids for 1–22 and the X. For PB2 the nomenclature is given entirely in relation to the number of chromatids where the normal haploid set is one chromatid for 1–22 and X. These terms however, are also appropriate to describe the respective oocytes. It should also be noted that although PBs are analyzed, the deduced result of the oocyte is needed to make a diagnosis, i.e., whether or not the oocyte is chromosomally normal so it can be fertilized for further development into an embryo.

Polar bodies 1 and 2 have different numbers of chromatids, and the microarray/NGS profiles regarding copy number should be interpreted accordingly.

If shallow NGS is used, replace "**arr**" with "**sseq**" for normal and abnormal results. As the resolution of the shallow NGS is approximately 5–10 Mb, the normal result can be described, acknowledging the limitations of the technique.

Normal

arr cht(X,1–22)×2
: Normal result for polar body 1 using microarray.

arr cht(X,1–22)×1
: Normal result for polar body 2 using microarray.

sseq cht(X,1–22)×2
: Normal result for polar body 1 using shallow NGS. The limitations of the test are outlined in the interpretive text.

sseq cht(X,1–22)×1
: Normal result for polar body 2 using shallow NGS. The limitations of the test are outlined in the interpretive text.

Abnormal, aneuploidy PB1

The normal copy number is ×2 (2 chromatids) for PB1.

arr cht(X)×1
  Polar body 1 has loss of one chromatid X using microarray.

arr cht(X)×1,cht(5)×0
  Polar body 1 has loss of one chromatid X and has lost both chromatids 5 using microarray. Note that the sex chromatid abnormality is described first.

arr cht(4)×0,cht(17)×4,cht(18)×0,cht(22)×3
  Polar body 1 has loss of both chromatids 4 and 18, gain of two chromatids 17, and gain of one chromatid 22 using microarray.

arr cht(4,14)×0,cht(17)×4,cht(22)×3
  Polar body 1 has loss of both chromatids 4 and 14, gain of two chromatids 17, and gain of one chromatid 22 using microarray.

arr cht(8,19)×3
  Polar body 1 has gain of one chromatid 8 and a gain of one chromatid 19 using microarray. There are three chromatids in total for each aneuploidy. Note that parentheses can be used to group abnormalities of the same copy number.

arr cht(18,21)×3
  Polar body 1 has gain of a chromatid 18 and a gain of a chromatid 21 using microarray.

Abnormal, aneuploidy PB2

The normal copy number is ×1 (1 chromatid) for PB2.

arr cht(X)×0
  Polar body 2 has loss of the chromatid X using microarray.

arr cht(1)×0,cht(19)×2
  Polar body 2 has loss of the chromatid 1 and a gain of one chromatid 19 using microarray.

arr cht(13)×2,cht(21)×0
  Polar body 2 has gain of one chromatid 13 and a loss of the chromatid 21 using microarray.

arr cht(13,18,21)×2
  Polar body 2 has gain of one chromatid each for 13, 18 and 21 using microarray.

arr cht(13,14)×2,(18)×0,(21)×2
  Polar body 2 has gain of one chromatid each for 13, 14 and 21, and has lost the chromatid 18 using microarray.

Abnormal, structural PB1

arr[GRCh37] cht1p36.3p12(995,002_120,315,488)×1,cht1q12q44(133,475,616_249,197,778)×3,cht(21)×4
   Polar body 1 has chromatid loss of the region 1p36.3p12, chromatid gain of the region 1q12q44 plus gain of two chromatids for 21 using microarray.

arr[GRCh37] cht7p22.3q33(616,253_137,659,838)×0 dmat,cht13q12.11q31.1 (19,812,720_80,581,454)×4 dmat
   Polar body 1 has loss of two chromatids for region 7p22.3q33 and gain of two additional chromatids for the region 13q12.11q31.1 using microarray. The mother is a carrier of a balanced reciprocal translocation, 46,XX,t(7;13)(q33;q31.1).

arr[GRCh37] cht8p23.3q11.21(125,733_50,822,069)×3,cht(21)×0
   Polar body 1 has partial gain of one chromatid for the 8p23.3q11.21 region and loss of two chromatids for 21 by microarray.

Abnormal, structural PB2

arr[GRCh37] cht1p36.3p12(995,002_120,315,488)×2,cht1q12q44(133,475,616_249,197,778)×0,cht(21)×0
   Polar body 2 has chromatid gain of the region 1p36.3p12, chromatid loss of the region 1q12q44, and loss of the chromatid for 21 using microarray.

arr[GRCh37] cht7p22.3q33(616,253_137,659,838)×3,cht13q12.11q31.1(19,812,720_80,581,454)×0,cht13q31.1q34(80,581,454_113,320,015)×2
   Polar body 2 has gain of two chromatids for region 7p22.3q33, loss of the chromatid for the region 13q12.11q31.1 and gain of a chromatid for the region 13q31.1q34 using microarray.

arr[GRCh37] cht8p23.3q11.21(125,733_50,822,069)×0,cht(21)×3
   Polar body 2 has chromatid loss of the region 8p23.3q11.21 and gain of two chromatids for 21 using microarray.

15 Region-Specific Assays

15.1 Introduction

Several diagnostic technologies exist that can be used to quantify the number of copies of a particular locus. Multiplex ligation-dependent probe amplification (MLPA), quantitative fluorescent PCR, real-time PCR, non-invasive prenatal diagnosis (NIPD), and bead-based assays can all be used to determine the number of copies of a chromosome and/or chromosomal region. These assays have collectively been referred to in this chapter as region-specific assays (**rsa**), which can be applied also to small, targeted arrays that are limited to a number of regions that can be reasonably listed in the nomenclature. rsa cannot be used for screening technologies, e.g., NIPT. The written description should indicate the resolution and limitations of the test.

When a kit is used, the name of the kit can be designated if the genomic coordinates are not known; however, the greatest precision is achieved by providing nucleotide numbers. As is done for microarray, the span of the abnormal nucleotides, with or without commas, is separated by an underscore. When nucleotide numbers are used or exons are given, the specified genome build (e.g., [GRCh38]) must be placed within the nomenclature string. Although not presented in the nomenclature, the reference sequence should be presented in the written description. To indicate a mixed cell population, the proportion of cells with the abnormality can be estimated and included in brackets following the copy number. If the proportion of abnormal cells cannot be estimated, the copy number range may be given.

Normal chromosomes and aberrations are listed from lowest to highest chromosome; sex chromosomes should be listed first. The decision to list the normal loci included in the assay is at the discretion of the laboratory.

15.2 Examples of RSA Nomenclature for Normal and Aneuploidy

46,XX.rsa(X,13,18,21)×2
    Normal female karyotype and normal copy number of chromosomes 13, 18, 21, and X using a region-specific assay.

rsa(X,Y)×1,(13,18,21)×2
    Normal copy number of chromosomes 13, 18, 21, X, and Y in a male using a region-specific assay.

rsa(21)×3
rsa(X,Y)×1,(13,18)×2,(21)×3
    Abnormal copy number result for chromosome 21 showing a gain for the whole chromosome (trisomy) in a male using a region-specific assay. For clarity, the normal disomic states for chromosomes 13, 18, and the sex chromosomes, which were also tested, may be included in the nomenclature.

rsa(X)×2,(Y)×1,(13)×2,(18)×3,(21)×2
    Abnormal copy number result for chromosomes X and 18 showing a gain for the whole chromosome 18 (trisomy) in an XXY male using a region-specific assay. For clarity, the normal disomic states for chromosomes 13 and 21 may be included in the nomenclature.

rsa(X)×2,(13)×3,(18,21)×2
    Abnormal copy number result for chromosome 13 showing a gain for the whole chromosome (trisomy) in a female using a region-specific assay. For clarity, the normal disomic states for chromosomes 18, 21, and X may be included in the nomenclature.

rsa(X)×1,(13,18,21)×2
    Abnormal copy number showing a loss of one sex chromosome using a region-specific assay, consistent with monosomy X using a region-specific assay. For clarity, the normal disomic states for chromosomes 13, 18, and 21 may be included in the nomenclature.

rsa(X)×2,(Y)×1,(13,18,21)×3
    Abnormal copy number showing an additional X chromosome plus gain of chromosomes 13, 18, and 21 (three copies) in a male using a region-specific assay. This result may be indicative of triploidy.

rsa(X)×2,(Y)×1,(21)×3
    Abnormal copy number result showing an additional X chromosome in a male along with an abnormal copy number for chromosome 21 showing a gain for the whole chromosome (trisomy) using a region-specific assay.

rsa(X)×2,(13,18)×2,(21)×2~3
    Abnormal copy number result for chromosome 21 showing two to three copies for the whole chromosome 21 (mosaic trisomy) in a female using a region-specific assay. For clarity, the normal disomic states for chromosomes 13, 18, and X may be included in the nomenclature when tested.

rsa(X,13)×2,(18)×3[0.6],(21)×2
    Abnormal copy number result for chromosome 18 showing mosaicism for the whole chromosome 18 (mosaic trisomy) in a female. 60% of the cells have this gain using a region-specific assay. For clarity, the normal disomic states for chromosomes 13, 21 and X may be included in the nomenclature when tested.

15.3 Examples of RSA Nomenclature for Partial Gain or Loss

rsa[GRCh38] Xp21.1(32,448,538_32,472,228)×1
rsa[GRCh38] Xp21.1(32441314×2,32448538_32472228×1,32484970×2)
    Abnormal copy number result showing loss within the *DMD* gene by MLPA in a female. The detailed form shows that the next neighboring nucleotides that do not show a loss are 7,224 and 12,742 nucleotides away from the alteration.

rsa[GRCh38] Xp21.2p21.1(31,037,731_33,457,670)×1
Abnormal copy number result for Xp21.1 showing a 2.4-Mb loss within the *DMD* gene by relative haplotype dosage using NIPD.

rsa[GRCh38] 1p36.33(849,466_2,432,509)×1
Abnormal copy number result for 1p36.33 showing a loss using a region-specific assay.

rsa 1p36.33(P070-B3)×1
Abnormal copy number result for 1p36.33 showing a loss using the MLPA kit P070-B3. The kit number is given as the nucleotide coordinates are not known.

rsa 20q13.3(P070-B3)×1,22q13.3(P070-B3)×1
Abnormal copy number result for both 20q13.3 and 22q13.3 showing a loss of both subtelomeric regions as defined by the MLPA kit P070-B3. The kit number is given as the nucleotide coordinates are not known.

arr[GRCh38] 8p23.1(8,479,797_11,897,580)×1.rsa[GRCh38] 8p23.1(11,676,959_11,760,002)×1
Microarray analysis shows a deletion of 8p at sub-band 8p23.1 which was confirmed using a region-specific assay targeting the *GATA4* locus. The coordinates using the region-specific assay do not confirm the extent of the 8p deletion.

rsa[GRCh38] 4q32.2q35.1(163,146,681_183,022,312)×1
Abnormal copy number result for 4q32.2q35.1 showing a loss using a region-specific assay.

rsa 13q14.2(RB1,DLEU2)×1,13q34(LAMP1)×3
Abnormal copy number result showing a loss of the *RB1* and *DLEU2* genes and gain of the *LAMP1* gene using a region-specific assay. Genes are listed from pter to qter.

rsa 13q14.2(D13S319)×1,13q34(LAMP1)×1,17p13.1(TP53)×1
Abnormal CLL result showing a loss of the marker D13S319 plus the *LAMP1* gene and loss of the *TP53* gene using a region-specific assay. Loci and genes are listed in chromosomal order.

rsa 15q12(GABRB3)×1,16p11.2(LAT)×3
Abnormal copy number result showing a loss of the *GABRB3* gene and gain of the *LAT* gene using a region-specific assay.

rsa 15q11.2(UBE3A)×1,15q12(GABRB3)×3
Abnormal copy number result showing a loss of the *UBE3A* gene and a gain of the *GABRB3* gene using a region-specific assay.

rsa[GRCh38] 16q11.1q24.3(37,633,407_90,218,850)×3
Abnormal copy number result for 16q11.1q24.3 showing a gain using a region-specific assay.

rsa 22q11.2("kit name with version")×1
Abnormal copy number result showing a loss of 22q11.2 using an MLPA kit. The name of the kit can be inserted in the parentheses without the quotation marks.

rsa 22q11.21(HIRA)×1
Abnormal copy number result showing a loss of 22q11.21 using a region-specific assay targeting the *HIRA* locus.

rsa 22q11.21(MICAL3,HIRA)×1,22q11.21(MED15)×2,22q11.21(HIC2)×1,22q13.33 (ARSA)×1

> Abnormal copy number result showing a loss within 22q11.21 and 22q13.33 using a region-specific assay. Genes *MICAL3*, *HIRA*, *HIC2*, and *ARSA* show a loss. *MED15* has a normal copy number. The genes are listed from pter to qter within the same chromosome region according to the genome build. The normal copy number for *MED15* is given as this clarifies the extent of the abnormality.

rsa 22q11.21(CLDN5,GP1BB)×1,22q11.21(SNAP29,PPIL2)×2,22q11.22q11.23 (RTDR1)×1

> Abnormal copy number result showing a partial loss within 22q11 using a region-specific assay targeting several loci. The proximal (*CLDN5* and *GP1BB*) and distal (*RTDR1*) regions genes are deleted. *SNAP29* and *PPIL2* have a normal copy number.

15.4 Examples of RSA Nomenclature for Balanced Translocations or Fusion Genes

rsa(BCR::ABL1)neg

> Normal result using a region-specific assay to identify a *BCR-ABL1* translocation or juxtaposition.

rsa(BCR::ABL1)pos

> Abnormal result using a region-specific assay that shows a *BCR-ABL1* translocation or juxtaposition.

arr[GRCh38] (8)×3,9q34.11q34.3(129,300,000_140,273,252)×1.rsa(BCR::ABL1)pos, (CBFB::MYH11)pos

> Abnormal array result showing a gain of chromosome 8 and loss of chromosome band 9q34. In addition, two translocations were identified (*BCR-ABL1* and *CBFB-MYH11*) with region-specific assays.

16 Sequence-Based Nomenclature for Description of Chromosome Rearrangements

16.1 Introduction

Historically, the ISCN has covered the description of numerical and structural chromosome changes detected using a variety of traditional and molecular cytogenetic techniques while the Human Genome Variation Society (HGVS; varnomen.HGVS.org; den Dunnen et al., 2016) covered the description of changes at the nucleotide level. Both ISCN and HGVS have developed methods to describe variants detected by microarrays, PCR, MLPA and other technologies, e.g., to detect copy number variation (CNV). Given the increased use of sequencing technologies to characterize chromosomal abnormalities (Schluth-Bolard et al., 2013; Ordulu et al., 2014; Newman et al., 2015), and the standards already established by ISCN and HGVS, it has become evident that a combined ISCN and HGVS standard for the description of large chromosome rearrangements identified by sequence-based technologies is required. The method of combining ISCN-like description of chromosome rearrangements with HGVS-like nucleotide variant descriptions, developed jointly between the ISCN and HGVS, was initially introduced in ISCN (2016).

16.2 General Principles

Key aspects of the combined standard:
- Where details of the chromosomal structure described have been derived from sequencing techniques, both the ISCN-like description of chromosome aberrations and the HGVS-like nucleotide variant descriptions should be included for clarity and completeness. When only breakpoint information is available without the information sufficient to describe the nature of the structural rearrangement, HGVS standards should be used alone. For description of large structural variation (see Section 16.3) the ISCN-like portion of the description appears first and can be modeled using the ISCN short or detailed form (see Chapter 9). Where several aberrations are described in one result, the short form may be used for some aberrations and the detailed form for other aberrations. For description of large copy number variation (see Section 16.4) the ISCN-like portion of the description appears first and can be modeled using the ISCN short or detailed form description of microarray (see Chapter 14).
- The ISCN-like portion of the description begins with **seq** to indicate that the aberration was characterized by sequence-based technology. It must include the genome build in square brackets after **seq**. When observations of other techniques such as banded

chromosome analysis or microarrays are available, they are presented first and a period (.) precedes the sequencing nomenclature (**.seq**).
- The combined nomenclature uses the existing HGVS standards for the HGVS-like portion (see below and https://varnomen.hgvs.org; den Dunnen et al., 2016) along with additional recommendations outlined below for the description of an aberration.
- Each derivative chromosome is described in a separate line of the HGVS-like portion.
- The segment containing the centromere is the reference for the derivative chromosomes. In order to have consistent descriptions, the directionality of the derivative chromosome should be defined by the directionality of the chromosome segment that includes the centromere. The "pter" of a derivative chromosome may be the original "qter" of a chromosome involved in a rearrangement. If so, the derivative chromosomes should be described from their new pter to qter, starting with the derivative chromosome with the lowest number, and not necessarily the pter of the chromosome with the lowest number involved. Aberrations affecting sex chromosomes are listed first (X then Y) followed by those affecting autosomes in ascending numbers.
- Sex chromosomes are not listed if normal, but the sex may be indicated in the written description.
- The genomic reference sequence contains undefined nucleotides (N's) at the start and end of the chromosome (telomeres), making the use of specific nucleotide positions problematic. The beginning and end of a chromosome are therefore represented as **pter** and **qter**.
- Multiple breakpoints in one chromosome are described from pter to qter of the derivative chromosome, anchored in the positive strand.
- The presence of an additional sequence which is not attached to other chromosomal material (i.e., trisomy, marker or ring chromosome) is indicated by **sup** (supernumerary chromosome) in the HGVS-like portion.
- For all rearrangements, when the breakpoints have not been determined at the precise nucleotide level, both the maximal (a_d) and minimal extent (b_c) of the rearrangement are indicated; format (a_b)_(c_d) (ranges of uncertainty given in parenthesis). Unknown nucleotide positions should be indicated using a "?", e.g., g.(?_b)_(c_?)del. For deletions extending from a known nucleotide position (#) to an unknown position in the direction of the telomere the format "(pter)_#" or "#_(qter)" is used.
- Complex results can be presented in a tabular form with the ISCN-like portion in one column and the HGVS-like portion in another column.
- Commas may be used in nucleotide numbers to indicate thousands and millions.
- When large copy number variations are identified by sequencing, the nature of a rearrangement on a chromosomal level is not determined and should not be inferred since only copy number information is obtained. The affected chromosome bands and the span of the aberrant nucleotides are given in a manner analogous to microarray in the ISCN-like description. This is followed by the HGVS-like nucleotide variant description.

Briefly, existing HGVS standards include the following:
- An underscore indicates a range of nucleotides. For example, g.123_456del indicates a deletion of nucleotides 123 to 456.
- Uncertainty is indicated by enclosing the uncertainty with parentheses. Thus, g.(123_234)_(450_856)dup indicates that a duplication starts at an unknown position between nucleotides 123 and 234 and extends to an unknown position between nucleotides 450 and 856.
- In runs of identical nucleotides, the most 3′ nucleotide in the run is designated as variant. This is termed the HGVS "3′ rule." For example, g.4delT (not g.2delT or g.3delT) describes the change CTTTA to CTTA.

- Inverted sequences are described using **inv** (e.g., g.123_456inv).
- Molecular databases store variants based on genomic coordinates, not chromosomal banding patterns. ISCN recommends the use of the translation tables provided by NCBI to translate banding patterns in chromosome nucleotide positions:
  - hg19/GRCh37 http://hgdownload.cse.ucsc.edu/goldenPath/hg19/database/cytoBand.txt.gz
  - hg38/GRCh38 http://hgdownload.cse.ucsc.edu/goldenPath/hg38/database/cytoBand.txt.gz
- A **double colon (::)** is used to designate break point junctions creating a ring chromosome.
- Non-templated sequences (inserts of nucleotides that do not align locally to the reference chromosome sequence) are described using **ins** (e.g., insAAGTAC). When the inserted sequence is long, the sequence can be replaced with a reference sequence accession.version number (e.g., ins[NG_012232.1:g.4566_8781]) specifying specific nucleotides.
- The term **delins** is used to describe a sequence change where, compared to a reference sequence, nucleotides are replaced by other nucleotides (e.g., derivative chromosomes, whether part of balanced or unbalanced translocations).

To determine the location of the breakpoint, the general HGVS rule of maintaining the longest unchanged sequence applies (the 3′ rule).

Application of the 3′ rule to chromosome rearrangements:

```
 5'                              3'
  Chr2:  TCAGC ATC CGTTGG_cen_qter
  Chr18: CAGTT ATC TCTGCC_cen_qter

  der2:  CAGTT ATC CGTTGG_cen_qter
  der18: TCAGC ATC TCTGCC_cen_qter
```

In the example above, a translocation joins chromosome 2 and chromosome 18 (bold). According to the chromosome 2 and chromosome 18 reference sequences (first two lines), there are two options to align the breakpoints (dashed and solid vertical lines). The 3′ rule determines that, starting with the lowest chromosome number involved (here chromosome 2), the sequence should be aligned as far 3′ as possible, aligning with the solid vertical line. The correct description of the rearrangement is therefore:

  seq[GRCh38] t(2;18)(p25.3;p11.32)
  NC_000002.12:g.pter_8delins[NC_000018.10:g.pter_8]
  NC_000018.10:g.pter_8delins[NC_000002.12:g.pter_8]

NOTE: A 5′ alignment would give NC_000002.12:g.pter_5delins[NC_000018.10:g.pter_5] and NC_000018.10:g.pter_5delins[NC_000002.12:g.pter_5]

16.3 Examples of Sequence-Based Nomenclature for Description of Large Structural Variation

16.3.1 Deletions

seq[GRCh38] del(X)(q21.31q22.1)
NC_000023.11:g.89555676_100352080del

Based on genome build GRCh38, genomic reference sequence NC_000023.11, an interstitial deletion within the long arm of one X chromosome from band Xq21.31 to band Xq22.1 has been identified. The deletion includes the segment from nucleotide 89,555,676 to nucleotide 100,352,080, joining nucleotide 89,555,675 to 100,352,081.

seq[GRCh38] del(X)(q21.31q22.1),del(X)(q21.31q22.1)
NC_000023.11:g.[89555676_100352080del];[89555676_100352080del]
    In a female, based on genome build GRCh38, genomic reference sequence NC_000023.11, a homozygous interstitial deletion within the long arm of both X chromosomes from band Xq21.31 to band Xq22.1 has been identified. The deletions include the segment from nucleotide 89,555,676 to nucleotide 100,352,080. In the HGVS portion each allele is described between square brackets ([ ]), separated by a semicolon (;). In the ISCN portion the alleles are distinguished by underlining one of them. Alternatively the ISCN portion could be written: seq[GRCh38] del(X)(q21.31q22.1)×2.

seq[GRCh38] del(X)(q21.31q22.1),del(X)(q21.31q22.1)
NC_000023.11:g.[89555674_100352088del];[90111276_99878726del]
    In a female, based on genome build GRCh38, genomic reference sequence NC_000023.11, a compound heterozygous interstitial deletion within the long arms of both X chromosomes from band Xq21.31 to band Xq22.1 has been identified. The deletions include the segment from nucleotide 89,555,674 to 100,352,088 on one homologue, and from nucleotide 90,111,276 to 99,878,726 on the other homologue. In the HGVS portion each allele is described between square brackets ([ ]), separated by a semicolon (;).

16.3.2 Derivative Chromosomes

seq[GRCh38] der(2)t(2;11)(p25.1;p15.2)
NC_000002.12:g.pter_8247756delins[NC_000011.10:g.pter_15825272]
    A derivative chromosome 2 from a translocation between the short arms of chromosomes 2 and 11 with breakpoints at 2p25.1 (between nucleotides 8,247,756 and 8,247,757) and 11p15.2 (between nucleotides 15,825,272 and 15,825,273), based on genome build GRCh38.

seq[GRCh38] der(3)(3pter→3q25.32::8q24.21→8qter)
NC_000003.12:g.158573187_qterdelins[NC_000008.11:g.(128534000_128546000)_qter]
    A derivative chromosome 3 from a translocation between the long arms of chromosomes 3 (breakpoint between nucleotides 158,573,186 and 158,573,187) and chromosome 8 (breakpoint at an unknown position between nucleotides 128,534,000 and 128,546,000).

seq[GRCh38] der(4)ins(4;X)(q28.3;q22.2q21.31)
NC_000004.12:g.134850793_134850794ins[NC_000023.11:g.[89555676_89556011;
(89556012_1003519998);100351999_100352080]inv]
    An unbalanced interchromosomal insertion with X chromosome long arm material inserted into the long arm of chromosome 4. The inserted sequence from the X chromosome is inverted in orientation relative to the chromosome X reference sequence. Sequence data are from the site of the insertion (134,850,793 to 134,850,794) and from the inserted sequence from positions 89,555,676 to 89,556,011 and 100,351,999 to 100,352,080, i.e., at the junction breakpoints.

seq[GRCh38] der(5)t(5;10)(p13.3;q21.3)
NC_000005.10:g.pter_29658442delins[NC_000010.11:g.67539995_qterinv]
    Sequencing reveals a derivative chromosome 5 from a translocation between the short arm of chromosome 5 and the long arm of chromosome 10. The segment including pter to nucleotide 29,658,442 of chromosome 5 has been replaced by nucleotides 67,539,995 to qter from chromosome 10, which are present in an inverted orientation relative to the orientation of the reference sequence (the original sequence of chromosome 10).

seq[GRCh38] der(6)t(6;13)(q13.3;q31.1),der(13)t(6;13)(q13.3;q31.1)inv(6)(q14.3q14.3)
NC_000006.12:g.85897871_qterdelins[A;NC_000013.11:g.80659609_qter]
NC_000013.11:g.80659607_qterdelins[NC_000006.12:g.[85897899_85900540inv;
85900541_86488291;93909933_qter]]
> A complex rearrangement between chromosomes 6 and 13. There is a single bp (an A) inserted at the breakpoint on the derivative chromosome 6. There is a 2-bp deletion of chromosome 13 material (80,659,607 to 80,659,608) and a 28-bp deletion (85,897,871 to 85,897,898) of chromosome 6 material at the breakpoint. The derivative chromosome 13 has a 2.6-kb inversion (85,897,899 to 85,900,540) along with a 7.4-Mb deletion (86,488,292 to 93,909,932) of chromosome 6 material.

seq[GRCh37] der(6)(6pter→6q14.1::21q22.12→qter),der(12)(6qter→q23.2::12p13.2→qter),der(21)(21pter→q22.12::12p13.2→12pter)
NC_000006.11:g.79,662,191_qterdelins[NC_000021.8:g.36,414,387_qter]
NC_000012.11:g.pter_12,031,320delins[NC_000006.11:g.132,835,666_qterinv]
NC_000021.8:g.36,411,882_qterdelins[NC_000012.11:g.12,031,429_pterinv]
> A rearrangement between chromosomes 6, 12, and 21 resulting in three derivative chromosomes. There are 53-Mb, 109-bp and 2.5-kb deletions of the chromosome 6, 12, and 21 sequences, respectively, evident from the nucleotide coordinates for the two breakpoints given for each chromosome.

16.3.3 Duplications

seq[GRCh38] dup(8)(q24.21q24.21)
NC_000008.11:g.128746677_128749160dup
> The sequenced breakpoint of a 2.4-kb duplication within chromosome 8, reference sequence NC_000008.11 band q24.21 including nucleotides 128,746,677 to 128,749,160 based on genome build GRCh38. The orientation of the duplicated segment is the same orientation as the reference sequence.

seq[GRCh38] dup(8)(q24.21q24.21)
NC_000008.11:g.128746676_12874677ins128746677_128749160inv
> The insertion of a 2.4-kb duplicated sequence within chromosome 8, reference sequence NC_000008.11, band q24.21, between nucleotides 128,746,676 and 12,874,677 including nucleotides 128,746,677 to 128,749,160 based on genome build GRCh38. The orientation of the inserted (duplicated) segment is inverted relative to the reference sequence.

seq[GRCh38] dup(8)(q24.21q24.21)
NC_000008.11:g.128749160_128749161ins128746677_128749160inv
> The insertion of a 2.4-kb duplicated sequence within chromosome 8, reference sequence NC_000008.11, band q24.21, between nucleotides 128,749,160 and 128,749,161 including nucleotides 128,746,677 to 128,749,160 based on genome build GRCh38. The orientation of the inserted (duplicated) segment is inverted relative to the reference sequence.

16.3.4 Insertions

seq[GRCh38] ins(4;X)(q28.3;q21.31q22.2)
NC_000023.11:g.89555676_100352080del
NC_000004.12:g.134850793_134850794ins[NC_000023.11:g.89555676_100352080]
> A balanced interchromosomal insertion of chromosome X long arm material (nucleotides 89,555,676 to 100,352,080) into the long arm of chromosome 4 (between nucleotides 134,850,793 and 134,850,794). The inserted sequence from the X chromosome is in the same orientation as the reference sequence.

seq[GRCh38] ins(4;X)(q28.3;q22.2q21.31)
NC_000023.11:g.89555676_100352080del
NC_000004.12:g.134850793_134850794ins[NC_000023.11:g.89555676_100352080inv]
    A balanced interchromosomal insertion of chromosome X long arm material (nucleotides 89,555,676 to 100,352,080) into the long arm of chromosome 4 (between nucleotides 134,850,793 and 134,850,794). The inserted sequence from the X chromosome is inverted in orientation relative to the reference sequence. Note that the aberration of the X chromosome is listed first in the HGVS-like portion.

16.3.5 Inversions

seq[GRCh38] inv(6)(pter→p25.3::q16.1→p25.3::q16.1→qter)
NC_000006.12:g.[776788_93191545inv;93191546T>C]
    Using the detailed form, a pericentric inversion in chromosome 6 (nucleotides 776,788 to 93,191,545) with a T to C nucleotide substitution at the breakpoint (nucleotide 93,191,546).

seq[GRCh38] inv(2)(pter→p22.3::q31.1→p22.3::q31.1→qter)dn
NC_000002.12:g.[32310435_32310710del;32310711_171827243inv;insG]
    Using the detailed form, a *de novo* pericentric inversion in chromosome 2 (nucleotides 32,310,711 to 171,827,243) with a 276-bp deletion (nucleotides 32,310,435 to 32,310,710) at the short arm breakpoint and a single bp insertion (a G) at the long arm breakpoint.

seq[GRCh38] inv(2)(p22.3q31.1)mat
NC_000002.12:g.[32310435_32310710delinsG;32310711_171827243inv]
    Using the short form, a maternally inherited pericentric inversion in chromosome 2 (nucleotides 32,310,711 to 171,827,243) with a 276-bp deletion (nucleotides 32,310,435 to 32,310,710) and a single bp insertion (a G) at the short arm breakpoint.

seq[GRCh38] inv(6)(p22.3p21.2)
NC_000006.12:g.[20000000_40000000inv;40000001T>C]
    Using the short form, a paracentric inversion in the short arm of chromosome 6 (nucleotides 20,000,000 to 40,000,000) with a single nucleotide substitution at the breakpoint (40,000,001 T to C).

16.3.6 Ring Chromosomes

seq[GRCh38] r(8)(p23.2q24.3)
NC_000008.11:g.pter_33000000del::140000000_qterdel
    A ring derived from chromosome 8 with breakpoints at band p23.2 and q24.3 joining nucleotide 3,300,001 to nucleotide 139,999,999, based on genome build GRCh38.

seq[GRCh38] +r(8)(p23.2q24.3)
NC_000008.11:g.[pter_33000000del::140000000_qterdel]sup
    A supernumerary ring derived from chromosome 8 with breakpoints at band p23.2 and q24.3 joining nucleotide 3,300,001 to nucleotide 139,999,999, based on genome build GRCh38.

16.3.7 Translocations

Translocations that appear balanced karyotypically often show imbalance by sequencing. The following examples use ISCN to describe the gross structural change to chromosomes as the result of a translocation that was identified by NGS. The nucleotide positions in the HGVS-like portion show the extent and nature of any loss or gain.

46,XX,t(2;11)(p24;p15.1).seq[GRCh38] t(2;11)(p25.1;p15.2)
NC_000002.12:g.pter_8247756delins[NC_000011.10:g.pter_15825272]
NC_000011.10:g.pter_5825272delins[NC_000002.12:g.pter_8247756]
    A translocation between the short arms of chromosomes 2 and 11. The breakpoints, at bands 2p24 and 11p15.1 by banding, were further defined by sequencing of bands 2p25.1 and 11p15.2. Based on genome build GRCh38, there is joining of chromosome 11 nucleotide 15,825,272 to chromosome 2 nucleotide 8,247,757 on the derivative chromosome 2, and joining of chromosome 2 nucleotide 8,247,756 to chromosome 11 nucleotide 15,825,273 on the derivative chromosome 11.

seq[GRCh38] t(9;9)(9qter→9q22.33::9p21.2→9qter;9pter→9q22.33::9p21.2→9pter)
NC_000009.12:g.pter_26393001delins102425452_qterinv
NC_000009.12:g.102425452_qterdelinspter_26393001inv
    A translocation between homologous chromosomes with breakpoints at 9p21.2 and 9q22.33.

seq[GRCh38] t(2;11)(q31.1;q22.3)
NC_000002.12:g.17450009_qterdelins[NC_000011.10:g.108111987_qter]
NC_000011.10:g.108111982_qterdelins[NC_000002.12:g.17450009_qter]
    A translocation between the long arms of chromosomes 2 and 11. There is a 5-bp deletion of chromosome 11 sequence evident from the nucleotide numbers given for the two chromosome 11 breakpoints (108,111,982 and 108,111,987).

seq[GRCh38] t(3;14)(14qter→14q12::3p22.2→3qter;14pter→14q12::3p22.2→3pter)
NC_000003.12:g.pter_36969141delins[CATTTGTTCAAATTTAGTTCAAATGA;
NC_000014.9:g.29745314_qterinv]
NC_000014.9:g.29745314_qterdelins[NC_000003.12:g.36969141_pterinv]
    A translocation between the short arm of chromosome 3 (between nucleotides 36,969,141 and 36,969,142) and the long arm of chromosome 14 (between nucleotides 29,745,313 and 29,745,314) with insertion of non-templated sequence (CATTTGTTCAAATTTAGTTCAAATGA) at the breakpoint on the derivative chromosome 3.

seq[GRCh37] t(12;21)(p13.2;q22.12)
NC_000012.11:g.pter_12,027,787delins[NC_000021.8:g.36,326,005_qterinv]
NC_000021.8:g.36,325,405_qterdelins[NC_000012.11:g.12,026,106_pterinv]
    A translocation between the short arm of chromosome 12 and the long arm of chromosome 21. There is a 1.6-kb deletion of chromosome 12 sequence evident from the nucleotide numbers given for the two chromosome 12 breakpoints (12,026,106 and 12,027,787), and a 600-bp deletion of chromosome 21 sequence evident from the nucleotide numbers given for the two chromosome 21 breakpoints (36,325,405 and 36,326,005).

16.4 Examples of Sequence-Based Nomenclature for Description of Large Copy Number Variation

Copy number variation (CNV) can be identified by next-generation sequencing (NGS) technologies, also known as massively parallel sequencing (MPS). In exome sequencing or targeted sequencing breakpoints in introns will not be identified precisely. The following examples demonstrate this breakpoint uncertainty.

If a CNV is identified by shallow NGS, **sseq** should be used in the ISCN description of the variant (see Chapter 14.2.8).

seq[GRCh37] 12q21.32q21.32(88928628×2,88939469_88939743×0,88973993×2)mat pat
NC_000012.11:g.[(88928629_88939469)_(88939743_88973992)del];[(88928629_88939469)_(88939743_88973992)del]

> An intragenic, homozygous loss is detected in the *KITLG* gene on chromosome 12 using CNV in exome data analysis. The breakpoints are uncertain due to the technical limitations of the method. Both parents are known heterozygous carriers of this loss.

seq[GRCh37] 1p22.3p22.3(86558071×2,86578191_87380894×1,87458685×2)dn,
1q31.2q32.2(193219034×2,196197301_200619865×1,200628117×2)dn,2q33.1q34
(202010213×2,202013624_209436863×1,210517832×2)dn
NC_000001.10:g.(86558072_86578191)_(87380894_87458684)del(;)(193219035_196197301)_(200619865_200628116)del
NC_000002.11:g.(202010214_202013624)_(209436863_210517831)del

> Three interstitial *de novo* losses are detected by CNV calling in exome data analysis: a loss of approximately 900 kb in 1p22.3, a loss of approximately 5 Mb in 1q31.2q32.2, and a loss of approximately 8 Mb in 2q33.1q34. It is not determined whether the chromosome 1 deletions are on the same allele (cis) or in trans, as indicated by (;) in the HGVS portion. The breakpoints are uncertain due to the technical limitations of the method.

seq[GRCh37]) Xp22.31q28(6302018×1,6968382_qter×2),Yp11.2q12(9384903×1,10316153_qter×0)
NC_000023.10:g.(6302019_6968381)_(qter)dup
NC_000024.9:g.(9384904_10316152)_(qter)del

> CNV calling in exome data analysis detected two copies of the X chromosome in a male patient, but only one copy of the approximately 6.3-Mb distal end of the short arm is present. There is also one copy of the distal end of the short arm of the Y chromosome; the rest of the Y chromosome is absent. The breakpoints are uncertain due to the technical limitations of the method. Follow-up testing using routine cytogenetic analysis and FISH is essential to confirm the structural nature of the imbalance since the karyotype of this man is most likely 46,X,der(X)t(X;Y)(p22.31;p11.2).

seq[GRCh37] 14q31.3q32.33(83159163×2,85994993_104647150×3,104710536×2),
14q32.33(106237752×2,106303366_qter×1)
NC_000014.8:g.(83159164_85994993)_(104647150_104710535)dup(;)(106237753_106303366)_(qter)del

> CNV calling in exome data analysis detected two imbalances in chromosome 14: an interstitial gain of approximately 18 Mb in 14q31.3q32.33 and a terminal loss of approximately 1 Mb in 14q32.33. A normal copy number ($n = 2$) is found for the region of approximately 1.5 Mb between these two imbalances. It is not proven that the gained chromosome 14 material is within chromosome 14 or inserted elsewhere in the genome; follow-up testing using routine cytogenetic analysis and FISH is essential to confirm the structural nature of the imbalance. Also, it is not proven that the aberrations are in cis or trans which is indicated by the use of (;) in the HGVS portion.

17 References

Al-Aish MS: Human chromosome morphology. I. Studies on normal chromosome characterization, classification and karyotyping. Can J Genet Cytol 11:370–381 (1969).

Baliakas P, Jeromin S, Iskas M, Puiggros A, Plevova K, Nguyen-Khac F, Davis Z, Rigolin GM, Visentin A, Xochelli A, et al: Cytogenetic complexity in chronic lymphocytic leukemia: definitions, associations, and clinical impact. Blood 133:1205–1216 (2019).

Caspersson T, Farber S, Foley GE, Kudynoski J, Modest EJ, Simonsson E, Wagh U, Zech L: Chemical differentiation along metaphase chromosomes. Exp Cell Res 49:219–222 (1968).

Caspersson T, Lomakka G, Zech L: The 24 fluorescence patterns of human metaphase chromosomes – distinguishing characters and variability. Hereditas 67:89–102 (1972).

Chicago Conference (1966): Standardization in Human Cytogenetics. Birth Defects: Original Article Series, Vol 2, No 2 (The National Foundation, New York 1966).

Chun K, Hagemeijer A, Iqbal A, Slovak ML: Implementation of standardized international karyotype scoring practices is needed to provide uniform and systematic evaluation for patients with myelodysplastic syndrome using IPSS criteria: An International Working Group on MDS Cytogenetics Study. Leukemia Res 34:160–165 (2010).

Cremer T, Landegent J, Bruckner A, Scholl HP, Schardin M, Hager HD, Devilee P, Pearson P, van der Ploeg M: Detection of chromosome aberrations in the human interphase nucleus by visualization of specific target DNAs with radioactive and non-radioactive *in situ* hybridization techniques: diagnosis of trisomy 18 with probe L1.84. Hum Genet 74:346–352 (1986).

den Dunnen JT, Dalgleish R, Maglott DR, Hart RK, Greenblatt MS, McGowan-Jordan J, Roux AF, Smith T, Antonarakis SE, Taschner PE: HGVS recommendations for the description of sequence variants: 2016 update. Hum Mutat 37:564–569 (2016).

Denver Conference (1960): A proposed standard system of nomenclature of human mitotic chromosomes. Lancet i:1063–1065 (1960).

Dutrillaux B: Obtention simultanée de plusieurs marquages chromosomiques sur les mêmes préparations, après traitement par le BrdU. Humangenetik 30:297–306 (1975).

Ford CE, Hamerton JL: The chromosomes of man. Nature 178:1020–1023 (1956).

Francke U: High-resolution ideograms of trypsin-Giemsa banded human chromosomes. Cytogenet Cell Genet 31:24–32 (1981).

Francke U: Digitized and differentially shaded human chromosome ideograms for genomic applications. Cytogenet Cell Genet 65:206–218 (1994).

Francke U, Oliver N: Quantitative analysis of high-resolution trypsin-Giemsa bands on human prometaphase chromosomes. Hum Genet 45:137–165 (1978).

Grimwade D, Hills RK, Moorman AV, Walker H, Chatters S, Goldstone AH, Wheatley K, Harrison CJ, Burnett AK, National Cancer Research Institute Adult Leukaemia Working Group: Refinement of cytogenetic classification in acute myeloid leukemia: determination of prognostic significance of rare recurring chromosomal abnormalities among 5876 younger adult patients treated in the United Kingdom Medical Research Council trials. Blood 116:345–365 (2010).

Guan XY, Meltzer PS, Trent JM: Rapid generation of whole chromosome painting probes (WCPs) by chromosome microdissection. Genomics 22:101–107 (1994).

Haase D, Stevenson KE, Neuberg D, Maciejewski JP, Nazha A, Sekeres MA, Ebert BL, Garcia-Manero G, Haferlach C, Haferlach T, et al.: *TP53* mutation status divides myelodysplastic syndromes with complex karyotypes into distinct prognostic subgroups. Leukemia 33:1747–1758 (2019).

ISCN (1978): An International System for Human Cytogenetic Nomenclature. Birth Defects: Original Article Series, Vol 14, No 8 (The National Foundation, New York 1978); also in Cytogenet Cell Genet 21:309–404 (1978).

ISCN (1981): An International System for Human Cytogenetic Nomenclature – High Resolution Banding. Birth Defects: Original Article Series, Vol 17, No 5 (March of Dimes Birth Defects Foundation, New York 1981); also in Cytogenet Cell Genet 31:1–23 (1981).

ISCN (1985): An International System for Human Cytogenetic Nomenclature, Harnden DG, Klinger HP (eds), Birth Defects: Original Article Series, Vol 21, No 1 (March of Dimes Birth Defects Foundation, New York 1985).

ISCN (1991): Guidelines for Cancer Cytogenetics, Supplement to An International System for Human Cytogenetic Nomenclature, Mitelman F (ed), (S Karger, Basel 1991).

ISCN (1995): An International System for Human Cytogenetic Nomenclature, Mitelman F (ed), (S Karger, Basel 1995).

ISCN (2005): An International System for Human Cytogenetic Nomenclature, Shaffer LG, Tommerup N (eds), (S Karger, Basel 2005).

ISCN (2009): An International System for Human Cytogenetic Nomenclature, Shaffer LG, Slovak ML, Campbell LJ (eds), (S Karger, Basel 2009).

ISCN (2013): An International System for Human Cytogenetic Nomenclature, Shaffer LG, McGowan-Jordan J, Schmid M (eds), (S Karger, Basel 2012).

ISCN 2016: An International System for Human Cytogenomic Nomenclature (2016); McGowan-Jordan J, Simons A, Schmid M (eds), (Karger, Basel 2016); also in Cytogenet Genome Res 149:1–140 (2016).

Jhanwar SC, Burns JP, Alonso ML, Hew W, Chaganti RSK: Mid-pachytene chromomere maps of human autosomes. Cytogenet Cell Genet 33:240–248 (1982).

Kallioniemi A, Kallioniemi OP, Sudar D, Rutovitz D, Gray JW, Waldman F, Pinkel D: Comparative genomic hybridization for molecular cytogenetic analysis of solid tumors. Science 258:818–821 (1992).

Landegent JE, Jansen in de Wal N, Dirks RW, Baao F, van der Ploeg M: Use of whole cosmid cloned genomic sequences for chromosomal localization by non-radioactive *in situ* hybridization. Hum Genet 77:366–370 (1987).

Levan A, Frega K, Sandberg AA: Nomenclature for centromeric position on chromosomes. Hereditas 52:201–220 (1964).

Lichter P, Cremer T, Borden J, Manuelidis L, Ward DC: Delineation of individual human chromosomes in metaphase and interphase cells by *in situ* suppression hybridization using recombinant DNA libraries. Hum Genet 80:224–234 (1988).

Lichter P, Tang CJ, Call K, Hermanson G, Evans GA, Housman D, Ward DC: High-resolution mapping of human chromosome 11 by *in situ* hybridization with cosmid clones. Science 247:64–69 (1990).

Liehr T, Claussen U, Starke H: Small supernumerary marker chromosomes (sSMC) in humans. Cytogenet Genome Res 107:55–67 (2004).

Liehr T, Starke H, Heller A, Kosyakova N, Mrasek K, Gross M, Karst C, Steinhaeuser U, Hunstig F, Fickelscher I, et al: Multicolor fluorescence *in situ* hybridization (FISH) applied to FISH-banding. Cytogenet Genome Res 114:240–244 (2006).

London Conference on the Normal Human Karyotype. Cytogenetics 2:264–268 (1963).

Magenis RE, Barton SJ: Delineation of human prometaphase paracentromeric regions using sequential GTG- and C-banding. Cytogenet Cell Genet 45:132–140 (1987).

Newman S, Hermetz KE, Weckselblatt B, Rudd K: Next-generation sequencing of duplication CNVs reveals that most are tandem and some create fusion genes at breakpoints. Am J Hum Genet 96:208–220 (2015).

Ordulu Z, Wong KE, Currall BB, Ivanov AR, Pereira SA, Gusella JF, Talkowski ME, Morton CC: Describing sequencing results of structural chromosome rearrangements with a suggested next-generation cytogenetic nomenclature. Am J Hum Genet 94:1–15 (2014).

Paris Conference (1971): Standardization in Human Cytogenetics. Birth Defects: Original Article Series, Vol 8, No 7 (The National Foundation, New York 1972); also in Cytogenetics 11:313–362 (1972).

Paris Conference (1971), Supplement (1975): Standardization in Human Cytogenetics. Birth Defects: Original Article Series, Vol 11, No 9 (The National Foundation, New York 1975); also in Cytogenet Cell Genet 15:201–238 (1975).

Parra I, Windle B: High resolution visual mapping of stretched DNA by fluorescent hybridization. Nat Genet 5:17–21 (1993).

Patau K: The identification of individual chromosomes, especially in man. Am J Hum Genet 12:250–276 (1960).

Pinkel D, Landegent J, Collins C, Fuscoe J, Segraves R, Lucas J, Gray JW: Fluorescence *in situ* hybridization with human chromosome-specific libraries: detection of trisomy 21 and translocations of chromosome 4. Proc Natl Acad Sci USA 85:9138–9142 (1988).

Schluth-Bolard C, Labalme A, Cordier M-P, Till M, Nadeau G, Tevissen H, Lesda G, Boutry-Kryza N, Rossignol S, Rocas D, et al: Breakpoint mapping by next generation sequencing reveals causative gene disruption in patients carrying apparently balanced chromosome rearrangements with intellectual deficiency and/or congenital malformations. J Med Genet 50:144–150 (2013).

Stephens PJ, Greenman CD, Fu B, Yang F, Bignell GR, Mudie LJ, Pleasance ED, Lau KW, Beare D, Stebbings LA, et al: Massive genomic rearrangement acquired in a single catastrophic event during cancer development. Cell 144:27–40 (2011).

Tjio JH, Levan A: The chromosome number of man. Hereditas 42:1–16 (1956).

Trask BJ: Fluorescence *in situ* hybridization: applications in cytogenetics and gene mapping. Trends Genet 7: 149–154 (1991).

Viegas-Pequignot E, Dutrillaux B: Une méthode simple pour obtenir des prophases et des prometaphases. Ann Genet 21:122–125 (1978).

Wiegant J, Kalle W, Mullenders L, Brookes S, Hoovers JM, Dauwerse JG, van Ommen GJ, Raap AK: High-resolution *in situ* hybridization using DNA halo preparations. Hum Mol Genet 1:587–591 (1992).

Wiegant J, Wiesmeijer CC, Hoovers JM, Schuuring E, d'Azzo A, Vrolijk J, Tanke HJ, Raap AK: Multiple and sensitive fluorescence *in situ* hybridization with rhodamine-, fluorescein-, and coumarin-labeled DNAs. Cytogenet Cell Genet 63:73–76 (1993).

Wyandt HE, Tonk VS (eds): Atlas of Human Chromosome Heteromorphisms (Springer, New York 2008).

Yunis JJ: High resolution of human chromosomes. Science 191:1268–1270 (1976).

Yunis JJ, Sawyer JR, Ball DW: The characterization of high-resolution G-banded chromosomes of man. Chromosoma 67:293–307 (1978).

18   Members of the ISCN Standing Committee

Laura K. Conlin
Division of Genomic Diagnostics
Department of Pathology and Laboratory Medicine
Children's Hospital of Philadelphia
Philadelphia, PA, USA

Johan T. den Dunnen
Human Genetics and Clinical Genetics
Leiden University Medical Center
Leiden, The Netherlands

Rosalind J. Hastings
GenQA, Women's Centre, John Radcliffe Hospital
Oxford University Hospitals NHS Foundation Trust
Oxford, UK

Jin-Yeong Han
Department of Laboratory Medicine
Dong-A University College of Medicine
Busan, South Korea

Nils Mandahl
Department of Clinical Genetics
University and Regional Laboratories Lund University
Lund, Sweden

Jean McGowan-Jordan (Chair)
Departments of Genetics, CHEO and of
Pathology and Laboratory Medicine, University of Ottawa
Ottawa, ON, Canada

Sarah Moore
Genetics and Molecular Pathology, SA Pathology
SA Genomics Health Alliance
Adelaide, SA, Australia

Cynthia C. Morton
Departments of Obstetrics and Gynecology and of Pathology
Brigham and Women's Hospital, Harvard Medical School, Boston,
and Broad Institute of MIT and Harvard, Cambridge, MA, USA
Manchester Centre for Audiology and Deafness,
University of Manchester, Manchester, UK

Acknowledgements

The Committee gratefully acknowledges Martina Guttenbach, University of Würzburg, Germany, for her copy editing and members of the cytogenetics community for their suggestions and examples.

Statement of Ethics

As a set of recommendations this publication is exempt from ethical committee approval.

Conflict of Interest Statement

The authors have no conflicts of interest to declare.

Funding Sources

The 2019 Committee meeting and ISCN (2020) publication were made possible by generous contributions from Karger Publishers.

Author Contributions

Dr. Jean McGowan-Jordan collected suggestions from the cytogenetics community, collated all recommendations of the Standing Committee and edited all chapters. Dr. Rosalind Hastings reorganized Chapters 13 and 15, developed the polar body nomenclature in Chapter 14, provided critical review of, and contributed ideas and content for several other chapters. Dr. Sarah Moore provided critical review of, and additional examples for Chapters 11, 13, 14 and 16, and contributed ideas and content for several other chapters. As an ISCN Standing Committee member and Chair of the Human Genome Variation Society (HGVS) Sequence Variant Description Working Group, Dr. Johan den Dunnen contributed extensively to revisions of Chapter 16.

19 Appendix

Diagrammatic representation of human chromosome bands as observed with the Q-, G-, and R-staining methods; centromeric regions are representative of Q-staining method only (Paris Conference, 1971).

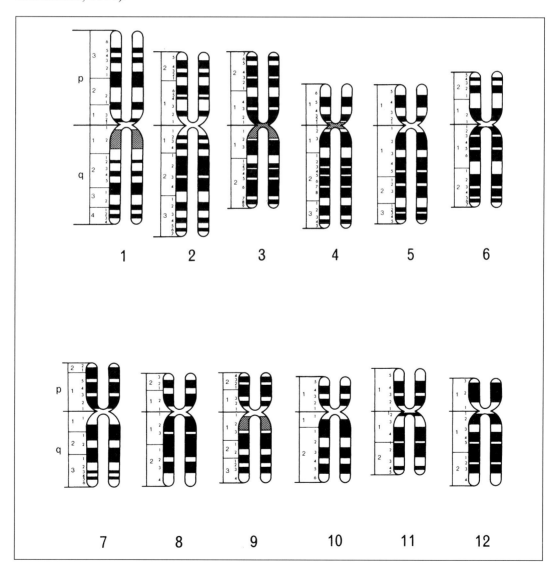

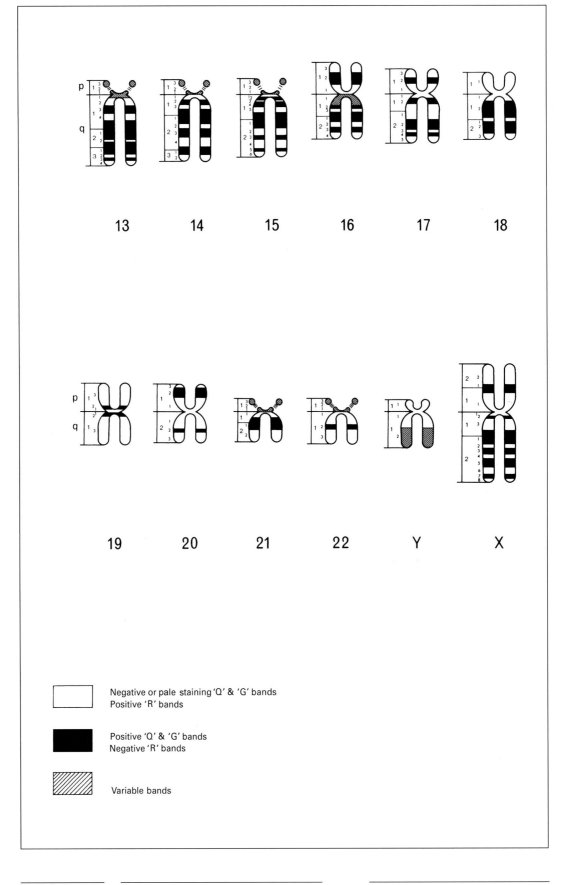

20 Index

The page numbers refer to Cytogenet Genome Res Vol. 160/7–8/2020

Abnormality, acquired   381, 396, 421, 435, 436, 463
   additional material   374, 379, 400
   alternative interpretation   375, 389
   autosomal   397
   clonal   417, 432, 437
   constitutional   374, 381, 396, 435, 463
   counting   437
   nonclonal   428, 432
   numerical   379, 390, 391, 395, 435, 437, 478
   order   378, 391, 466, 485
   sex chromosome   377, 379, 391, 396, 435, 466
   structural   377, 379, 382, 399, 437, 447, 479
Abnormally banded region   411
Acentric fragment   374, 414, 426
Achromatic lesion   425, 426
Acquired abnormality   381, 396, 421, 435, 436, 463
Additional material   374, 379, 400
Adjacent, segregation   385, 386
   signal   459
Alternative interpretation   375, 379, 389
Amplification   454, 455, 460
Anaphase   374, 438
Aneuploid   435, 478, 480
Angle brackets   374, 395
Arabic number   414, 417
Arm ratio   347
Array-based CGH   465
Arrow   374, 384
Association, telomeric   376, 417
Autosomal abnormality   397

Bacterial artificial chromosome (BAC)   465
Balanced rearrangement   378, 385, 418, 420, 437, 483
Band, definition   348, 349, 351
   designation   351, 379, 388
   high resolution   352
   interface   381, 410
   interval   389
Banding, high resolution   352
   molecular basis   354
   technique   348, 354

Bivalent   438, 439, 442
Brackets, angle   374, 395
   square   374, 377, 379, 429, 432, 455, 484
Break, chromatid   425
   chromosome   384, 426
   isochromatid   426
   isolocus   426
Break-apart probes   461
Breakpoint, designation   377, 382, 485, 486
   uncertainty   388, 485, 491

C-band   348, 353, 354, 438
Centromere, designation   351, 384
   fission   375, 409
   index   348
   premature division   376, 427
   suppressed   408
Chiasma, abbreviation   376, 439
   frequency   439
   interstitial   439
   location   439
Chimera   374, 379
Chromatid, break   425
   deletion   426
   exchange   425, 426
   gain   477
   gap   425
   interchange   425
   intrachange   425
   inversion   426
   loss   477
Chromatin, fiber   446, 463
   X   349
   Y   349
Chromoanasynthesis   374, 476
Chromosome, band   348, 349, 351, 384
   break   384, 426
   derivative   374, 384, 392, 402, 413, 485, 487
   dicentric   374, 407, 416, 422
   double minute   375, 414, 426
   exchange   375, 420, 426

gain   395, 428
gap   375, 426
group   347, 348
homologous   376, 406, 418
in situ hybridization   375, 447
iso   375, 412
isoderivative   375, 405
isodicentric   375, 407, 412
landmark   349, 355
loss   395, 428
marker   375, 384, 413, 427
meiotic   438
minute   375, 426
morphology   347
nomenclature   349
normal missing   379, 399
number   347, 389, 434
order in karyotype   347
paint   448, 464
Philadelphia   376, 403
premature condensation   376, 427
pseudodicentric   408
pseudotricentric   408
pulverization   376, 426
range   414, 432, 434
recombinant   376, 386, 402
region   349
ring   376, 415, 489
sub-band   351
tricentric   376, 416, 422
uncertainty   379, 388
unknown origin   400, 413
Chromothripsis   374, 426, 476
Clone, criteria   428
   definition   428
   evolution   428, 430
   frequency   429, 434
   normal   434
   order   430, 434
   polyploid   431
   presentation   430, 434
   related   429, 430
   size   429
   unrelated   434
Code, banding technique   354
Colon, double   374, 384, 486
   single   374, 384
Comma   374, 378, 447, 455, 465, 480, 485
Comparative genomic hybridization   465
Complete exchange   425
Complex, array results   466, 476
   interchange   425
   rearrangement   374, 381, 383, 419, 476
Composite karyotype   374, 432
Connected signal   374, 459
Constitutional abnormality   396, 435, 463
Constitutional karyotype   435
Copy number   455, 463, 466, 469, 472, 476, 478, 480
Copy number variation (CNV)   485, 491

Decimal point   351, 374
Definition, band   348, 349, 351
   landmark   349, 355
   region   349
   sub-band   351
Deletion   374, 401, 485, 486
   chromatid   426
   interstitial   401
   terminal   401
De novo   375, 381, 469
Derivative chromosome   374, 384, 392, 402, 413, 485, 487
Designation, band   351, 379, 388
   breakpoint   377, 382, 485, 486
   centromere   351, 384
   locus   447
   region   350, 351
   sub-band   351
Detailed form   382, 384
Diakinesis   374, 439
Dicentric chromosome   374, 405, 407, 416, 422
Dictyotene   375, 439
Diminished   374, 463
Diploid   434
Diplotene   375, 439
Disomy, uniparental   376, 474
Distal   375, 439
DNA, fiber   463
   gain   466, 481
   loss   466, 481
Donor   378, 380, 382, 458
Double minutes   375, 414, 426
Dual fusion probes   461
Duplication   375, 409, 488

Endoreduplication   380
Enhanced   375, 454, 463
Equal sign   375, 439
Euploid   435
Exchange, asymmetrical   425
   chromatid   425, 426
   chromosome   426
   complete   425
   incomplete   425
   sister chromatid   376, 425
   symmetrical   425

Fiber, chromatin   446, 463
   DNA   446, 463
   extended   463
FISH, see *In situ* hybridization
Fission, centric   375, 409
Four-break rearrangement   383, 419
Fragile site   375, 394, 410
Fragment, acentric   374, 414, 426
Fusion, centric   420
   genes   446, 483
   probes   461

Gap, chromatid 425
    chromosome 426
    isochromatid 426
    isolocus 426
G-band 348, 350, 354, 356, 370
Genome build 382, 466, 480, 484

Haploid 352, 356, 434
Heptaploid 435
Heterochromatin 348, 375, 393
Heterogeneity 432
Heteromorphism 352
Heterozygosity 375, 474
Hexaploid 435
High resolution banding 352
History of ISCN 341
Homogeneously staining region 375, 410
Homologous chromosomes 379, 391, 406, 418
    single underlining 379, 406, 418
Homozygosity 375, 474
Human Genome Variation Society (HGVS) 484, 485
Hybridization, comparative genomic 465
    in situ 375, 446
Hyperdiploid, -triploid, -tetraploid 434, 435
Hyphen 375, 446
Hypodiploid, -triploid, -tetraploid 434, 435

Idem 375, 429, 430
Identification, questionable 376, 388
Idiogram 347, 356, 442
Incomplete, exchange 425
    karyotype 375, 390
Inherited 375, 381, 469
Insertion 375, 383, 411, 488
In situ hybridization, extended fiber 463
    fluorescence (FISH) 446
    interphase 375, 455
    metaphase 447, 454
    nuclear 375, 455
    prophase 447
    reverse 463
Interchange, chromatid 425
Interpretation, alternative 375, 389
Interstitial, chiasma 439
    deletion 401
Interval 389
Intrachange, chromatid 425
Inversion 375, 412, 489
    chromatid 426
    paracentric 412
    pericentric 387, 412
Isochromatid, break 426
    gap 426
Isochromosome 375, 412
Isoderivative chromosome 375, 405
Isodicentric chromosome 375, 407, 412
Isolocus, break 426
    gap 426

Jumping translocation 422

Karyotype, composite 374, 432
    constitutional 435
    definition 347
    designation 378
    incomplete 375, 390
    normal 378
Karyotypic heterogeneity 432

Landmark 349, 355, 442
Leptotene 375, 439
Level of ploidy 380, 395, 434
List of abbreviations 354, 374, 439
Locus designation 447
Loss of heterozygosity 474, 475

Marker chromosome 375, 384, 392, 413, 427, 485
Maternal origin 375, 381, 386, 469
Medial 375, 439
Meiotic chromosomes 438
Microarray 465
    nomenclature 466
Minus sign 375, 379, 393, 395, 430, 438, 447
Minute 426
    double 375, 392, 414, 426
Modal number 434
Mosaicism 379, 452, 466
Multi-color chromosome painting 464
Multiple copies 375, 379, 414, 423, 453
Multiple techniques 470
Multiplex ligation-dependent probe amplification (MLPA) 480
Multiplication sign 375, 379, 414, 423, 447, 455
Multivalent 439

Near-diploid 395, 434
Near-haploid 395, 434
Neocentromere 375, 415
Neoplasia 428
Next-generation sequencing 466, 491
Nomenclature, chromosome band 349
    meiotic 439
    microarray 465, 466
    region-specific assay 480, 481, 483
    sequence-based 484, 486, 491
    SNP array 474
Nonclonal aberration 428
Non-invasive prenatal diagnosis (NIPD) 480
Normal karyotype 378
Nucleolus organizing regions (NORs) 348, 352
Number of cells, designation 379, 429
Numeral, Arabic 414, 417
    Roman 376, 438
Numerical abnormality 379, 390, 391, 395, 435, 437

Octaploid   435
Oligonucleotides   465
Oogonial metaphase   375, 439
Or   375, 379, 381, 389
Order, abnormalities   378, 391, 466, 485
   clones   430, 434
   karyotype   391

Pachytene   375, 439, 441, 442
   diagram   385, 387
Paint, partial chromosome   376, 464
   whole chromosome   376, 448
Paracentric inversion   412
Parentheses   376, 378, 382, 384, 438, 447, 455, 485
Partial chromosome paint   376, 464
Paternal origin   376, 381, 469
Pentaploid   435
Pericentric inversion   387, 412
Period   376, 446, 447, 455, 470, 485
Philadelphia chromosome   376, 403
Ploidy level   380, 395, 434
Plus sign, double   376, 447
   single   376, 379, 393, 395, 413, 417, 430, 438, 439, 447
Polar bodies, abnormal   478, 479
   normal   477
Polyploid   380, 431, 435
Premature centromere division   376, 427
Premature chromosome condensation   376, 427
Probes, break-apart   461
   dual fusion   461
   single fusion   461
   tricolor   462
Proximal   376, 439
Pseudodicentric, -tricentric   408
Pseudodiploid, -triploid   435
Pulverization   376, 426

Q-band   348, 349, 354
Quadriradial   376, 425
Quadrivalent   438
Quadruplication   376, 415
Questionable identification   376, 388
Question mark   376, 379, 381, 388

R-band   348, 351, 354, 372
Rearrangement, balanced   378, 385, 418, 420, 437, 483
   complex   374, 381, 383, 419, 476
   four-break   383, 419
   three-break   382, 419
   two-break   382, 418
Recipient   380, 383, 458
Reciprocal translocation   385, 386, 418
Recombinant chromosome   376, 386, 387, 402
Region, abnormally banded   411
   definition   349
   homogeneously staining   375, 410
Region-specific assay   480
Ring chromosome   376, 415, 489

   dicentric   416
   monocentric   416
   tricentric   416
Robertsonian translocation   376, 421
Roman numeral   376, 438
3′ Rule   485, 486

Satellite   348, 376, 393
Satellite stalk   376, 393
Semicolon   376, 378, 382, 405, 415, 438
Separated signal   376, 460
Sequencing   376, 485, 490, 491
Sex, chromatin   349
   chromosome abnormality   377, 379, 391, 396, 435, 466
Short form   382, 454, 455, 458, 465, 484
Sideline   376, 430
Sign
   equal   375, 439
   minus   375, 379, 393, 395, 430, 438, 447
   multiplication   375, 379, 414, 423, 447, 455
   plus   376, 379, 393, 395, 413, 417, 430, 438, 439, 447
Signal, adjacent   459
   amplified   374, 454, 460
   connected   374, 459
   intensity   454, 463
   number   447, 454, 455
   position   455, 458
   separated   376, 459
Signal pattern, abnormal   448, 452, 456
   chimeric   452
   normal   447, 456
Single fusion probes   461
Single nucleotide polymorphism (SNP)   465, 474
Sister chromatid exchange   376, 425, 426
Slant line, double   376, 380, 458
   single   376, 379, 428, 447, 455
SNP-array   474
Spermatogonial metaphase   376, 439
Square brackets   374, 377, 379, 429, 432, 455, 484
Stemline   376, 430
Sub-band   351
Subclone   428, 430
Subtelomeric region   376, 454
Suppressed centromere   408
Symbols, list   354, 374, 439

T-band   348
Telomeric association   376, 417
Terminal   376, 384, 439
Terminal deletion   401
Tetraploid   395, 434
Three-break rearrangement   382, 419
Tilde   376, 379, 389, 472, 475
Translocation   376, 379, 383, 418
   balanced   378, 418, 483

    complex   419
    jumping   422
    reciprocal   385, 386, 418
    Robertsonian   376, 421
    segregation   386
    whole-arm   420
Tricentric chromosome   376, 422
Tricolor probes   462
Triplication   423
Triploid   434
Triradial   376, 425
Trivalent   438
Two-break rearrangement   382, 418

Uncertainty, breakpoint localization   388, 485, 491
    chromosome number   389
    copy number   472
    nucleotides   485

Underlining   376, 379, 406, 418
Underscore   376, 465, 466, 480, 485
Uniparental disomy   376, 474
Univalent   438
Unknown material   400, 401
Unrelated clones   434

Variable chromosome region   347, 353, 376, 393

Whole-arm translocation   420
Whole chromosome paint   376, 448

X-chromatin   349

Y-chromatin   349

Zygotene   376, 439

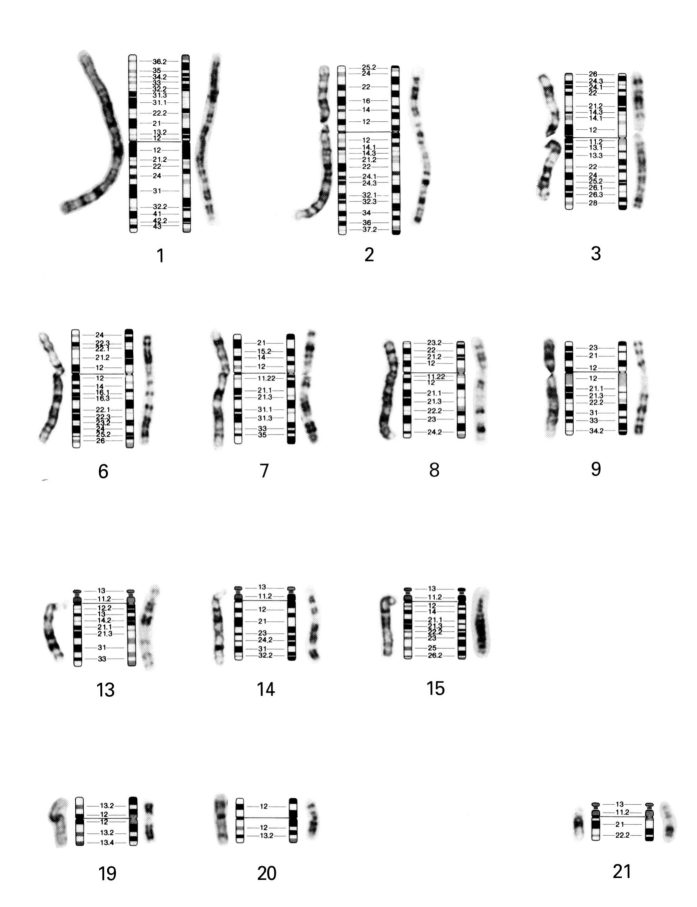

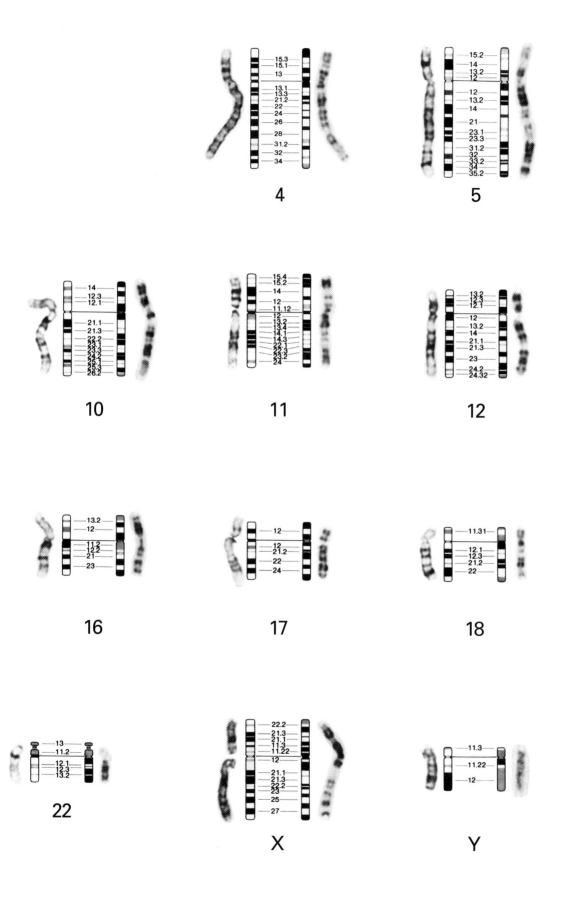

The Normal Human Karyotype G- and R-bands

Photographs of G- and R-banded human metaphase chromosomes and their diagrammatic representations (approximately 550-band stage). The diagrams are identical in the position and width of bands to those of the basic diagram (Fig. 5). For the G-band diagram (left) the G-positive bands have been shaded to match the intensity of the chromosomal bands in the photographs. In the case of the R-band diagram (right) the R-positive bands have been shaded to match the photographs. In both cases the negative bands are uniformly white. For convenience and clarity, only the G-positive bands are numbered. For the full numbering refer to Fig. 5. (Modified from ISCN 1985).

The G-banded photographs are taken, with permission from Francke, Cytogenet Cell Genet 31:24 (1981) and the R-banded photographs were provided by Dr. M. Prieur, with the assistance of her technicians, in the laboratory of Professor M. Vekemans, Hôpital Necker, Enfants Malades, Paris.